OFFICIAL SQA PAST PAPERS WITH ANSWERS

INTERMEDIATE 2 | UNITS 1, 2 & APPLICATIONS

MATHEMATICS 2008-2012

First exam published in 2008.
Published by Bright Red Publishing Ltd, 6 Stafford Street, Edinburgh EH3 7AU
tel: 0131 220 5804 fax: 0131 220 6710 info@brightredpublishing.co.uk www.brightredpublishing.co.uk

ISBN 978-1-84948-276-9

A CIP Catalogue record for this book is available from the British Library.

Bright Red Publishing is grateful to the copyright holders, as credited on the final page of the Question Section, for permission to use their material. Every effort has been made to trace the copyright holders and to obtain their permission for the use of copyright material. Bright Red Publishing will be happy to receive information allowing us to rectify any error or omission in future editions.

X101/202

NATIONAL QUALIFICATIONS 2008

TUESDAY, 20 MAY
1.00 PM – 1.45 PM

MATHEMATICS
INTERMEDIATE 2
Units 1, 2 and
Applications of Mathematics
Paper 1
(Non-calculator)

Read carefully

1 **You may NOT use a calculator.**

2 Full credit will be given only where the solution contains appropriate working.

3 Square-ruled paper is provided.

LI X101/202 6/9370

FORMULAE LIST

Sine rule: $\frac{a}{\sin A} = \frac{b}{\sin B} = \frac{c}{\sin C}$

Cosine rule: $a^2 = b^2 + c^2 - 2bc \cos A$ or $\cos A = \frac{b^2 + c^2 - a^2}{2bc}$

Area of a triangle: Area $= \frac{1}{2}ab \sin C$

Volume of a sphere: Volume $= \frac{4}{3}\pi r^3$

Volume of a cone: Volume $= \frac{1}{3}\pi r^2 h$

Volume of a cylinder: Volume $= \pi r^2 h$

Standard deviation: $s = \sqrt{\frac{\sum(x - \bar{x})^2}{n-1}} = \sqrt{\frac{\sum x^2 - (\sum x)^2/n}{n-1}}$, where n is the sample size.

Marks

ALL questions should be attempted.

1. A straight line has equation $y = 4x + 5$.
State the gradient of this line. **1**

2. Multiply out the brackets and collect like terms.

$$(3x + 2)(x - 5) + 8x$$ **3**

3. The stem and leaf diagram shows the number of points gained by the football teams in the Premiership League in a season.

Stem	Leaves
3	3 3 3 9
4	1 4 5 5 7 8
5	0 2 3 3 6 6
6	0
7	5 9
8	
9	0

n = 20   4 | 1 represents 41 points

(*a*) Arsenal finished 1st in the Premiership with 90 points.
In what position did Southampton finish if they gained 47 points? **1**

(*b*) What is the probability that a team chosen at random scored less than 44 points? **1**

4. (*a*) Factorise

$$x^2 - y^2.$$ **1**

(*b*) Hence, or otherwise, find the value of

$$9{\cdot}3^2 - 0{\cdot}7^2.$$ **2**

[Turn over

Marks

5. In a survey, the number of books carried by each girl in a group of students was recorded.

The results are shown in the frequency table below.

Number of books	*Frequency*
0	1
1	2
2	3
3	5
4	5
5	6
6	2
7	1

(*a*) Copy this frequency table and add a cumulative frequency column. **1**

(*b*) For this data, find:

(i) the median; **1**

(ii) the lower quartile; **1**

(iii) the upper quartile. **1**

(*c*) Calculate the semi-interquartile range. **1**

(*d*) In the same survey, the number of books carried by each boy was also recorded.

The semi-interquartile range was 0·75.

Make an appropriate comment comparing the distribution of data for the girls and the boys. **1**

6. Triangle PQR is shown below.

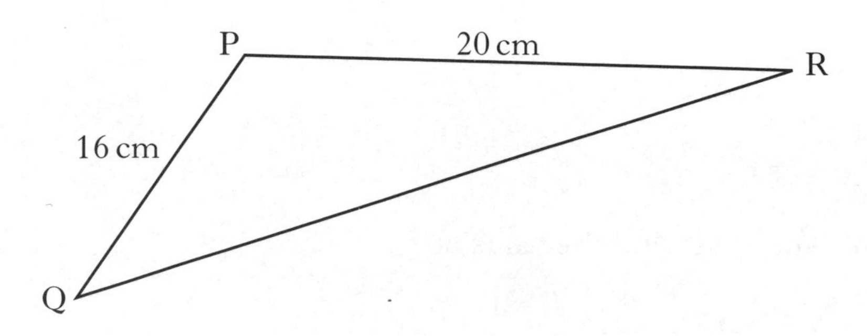

If $\sin P = \frac{1}{4}$, calculate the area of triangle PQR. **2**

Marks

7.

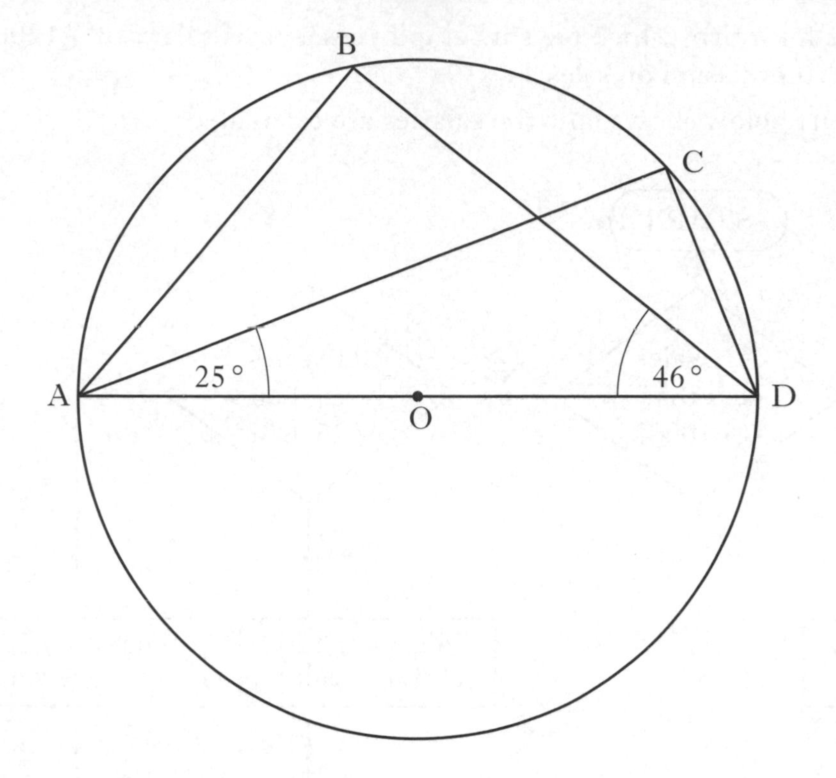

AD is a diameter of a circle, centre O.

B and C are points on the circumference of the circle.

Angle CAD = 25 °.

Angle BDA = 46 °.

Calculate the size of angle BAC. **3**

8. A network diagram is shown below.

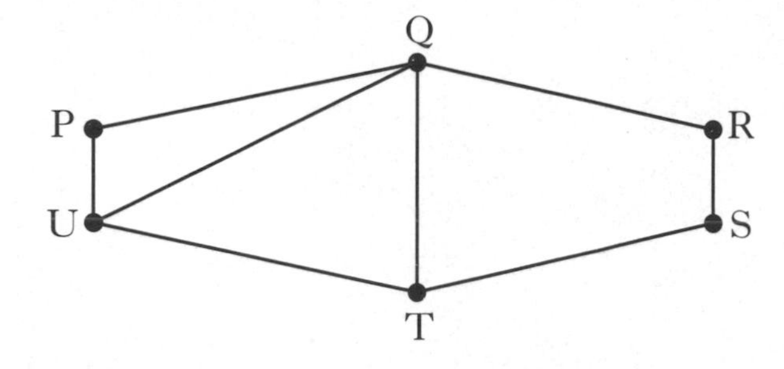

Write down the letters which represent the **odd** nodes. **1**

[Turn over

Marks

9. Jamie works for a firm which pays its employees a basic salary of £1200 per month plus commission on sales.

The flowchart below shows how the salaries are calculated.

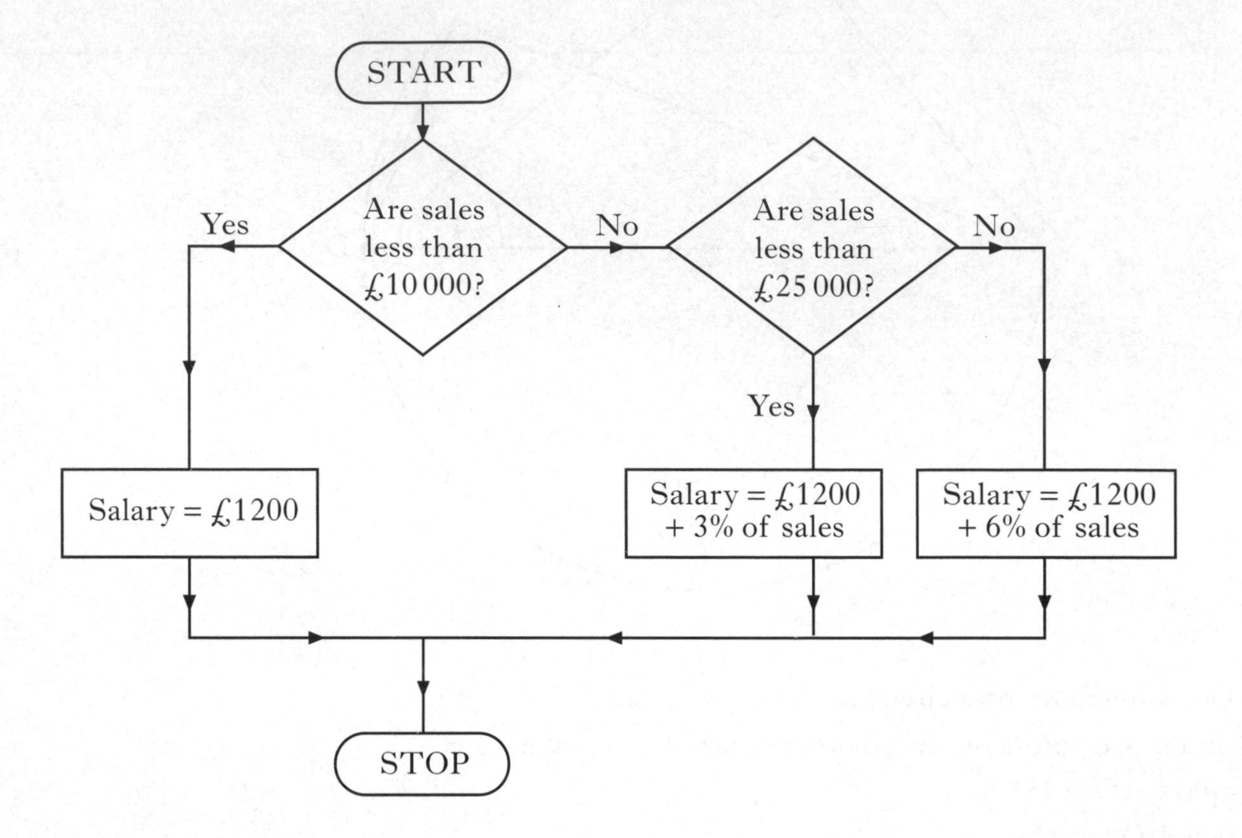

One month Jamie's sales are £40 000.

Calculate his salary for that month. **2**

Marks

10. A group of students was asked how many hours they spend studying each week. The histogram below shows the results of the survey.

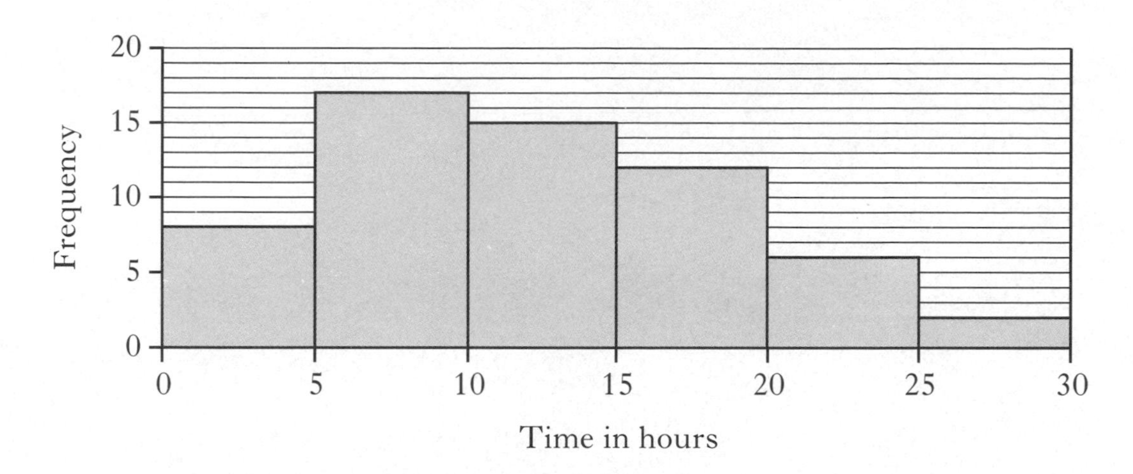

The **same** group of students was asked how many hours of television they watch each week.

The results of the survey are shown in the table below.

Time (h hours)	Frequency
$0 \leq h < 5$	1
$5 \leq h < 10$	4
$10 \leq h < 15$	9
$15 \leq h < 20$	20
$20 \leq h < 25$	14
$25 \leq h < 30$	12

(*a*) Using squared paper, draw a histogram to illustrate the results of this survey. **2**

(*b*) For the histogram you have drawn, estimate the mode to the nearest hour. **1**

(*c*) Compare the two histograms and comment. **1**

11. The sum of the terms of a sequence of numbers is given by the formula

$$S = \frac{a(r^n - 1)}{r - 1}.$$

Calculate S when $a = 3$, $r = 2$ and $n = 4$. **3**

[*END OF QUESTION PAPER*]

[BLANK PAGE]

X101/204

NATIONAL QUALIFICATIONS 2008

TUESDAY, 20 MAY
2.05 PM – 3.35 PM

MATHEMATICS
INTERMEDIATE 2
Units 1, 2 and
Applications of Mathematics
Paper 2

Read carefully

1 **Calculators may be used in this paper.**
2 Full credit will be given only where the solution contains appropriate working.
3 Square-ruled paper is provided.

FORMULAE LIST

Sine rule: $\frac{a}{\sin A} = \frac{b}{\sin B} = \frac{c}{\sin C}$

Cosine rule: $a^2 = b^2 + c^2 - 2bc \cos A$ or $\cos A = \frac{b^2 + c^2 - a^2}{2bc}$

Area of a triangle: Area $= \frac{1}{2}ab \sin C$

Volume of a sphere: Volume $= \frac{4}{3}\pi r^3$

Volume of a cone: Volume $= \frac{1}{3}\pi r^2 h$

Volume of a cylinder: Volume $= \pi r^2 h$

Standard deviation: $s = \sqrt{\frac{\sum(x-\bar{x})^2}{n-1}} = \sqrt{\frac{\sum x^2 - (\sum x)^2/n}{n-1}}$, where n is the sample size.

ALL questions should be attempted.

Marks

1. Calculate the **compound interest** earned when £50 000 is invested for 4 years at 4·5% per annum.

Give your answer to the nearest penny. **4**

2. Jim Reid keeps his washing in a basket. The basket is in the shape of a prism.

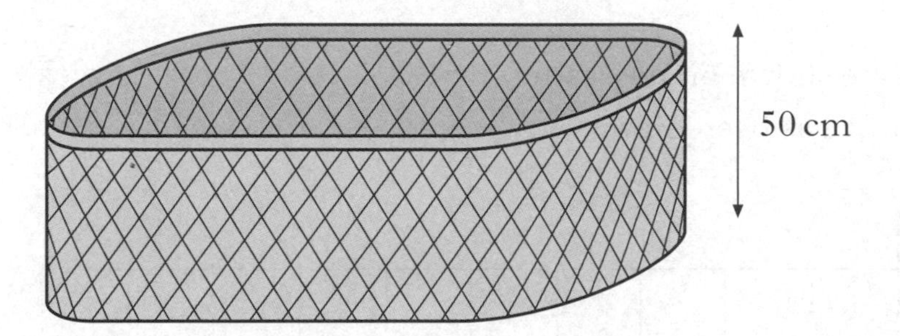

The height of the basket is 50 centimetres.

The cross section of the basket consists of a rectangle and two semi-circles with measurements as shown.

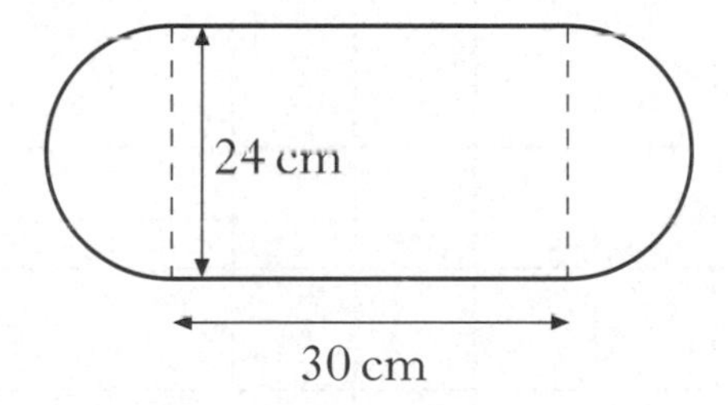

(*a*) Find the volume of the basket in cubic centimetres.

Give your answer correct to three significant figures. **4**

Jim keeps his ironing in a storage box which has a volume **half** that of the basket.

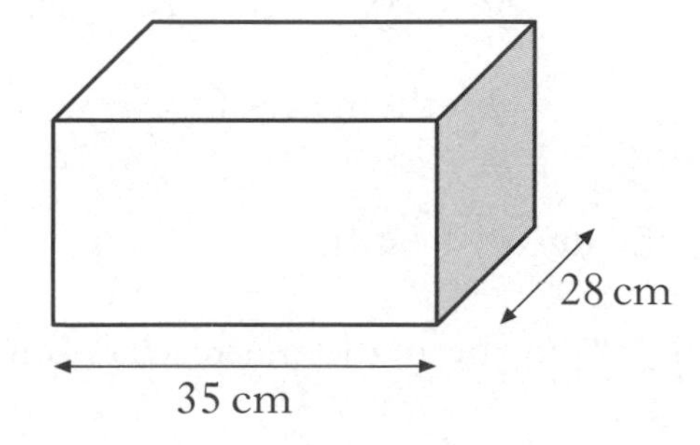

The storage box is in the shape of a cuboid, 35 centimetres long and 28 centimetres broad.

(*b*) Find the height of the storage box. **3**

Marks

3. The results for a group of students who sat tests in mathematics and physics are shown below.

Mathematics (%)	10	18	26	32	49
Physics (%)	25	35	30	40	41

(*a*) Calculate the standard deviation for the mathematics test. **4**

(*b*) The standard deviation for physics was 6·8.
Make an appropriate comment on the distribution of marks in the two tests. **1**

These marks are shown on the scattergraph below.
A line of best fit has been drawn.

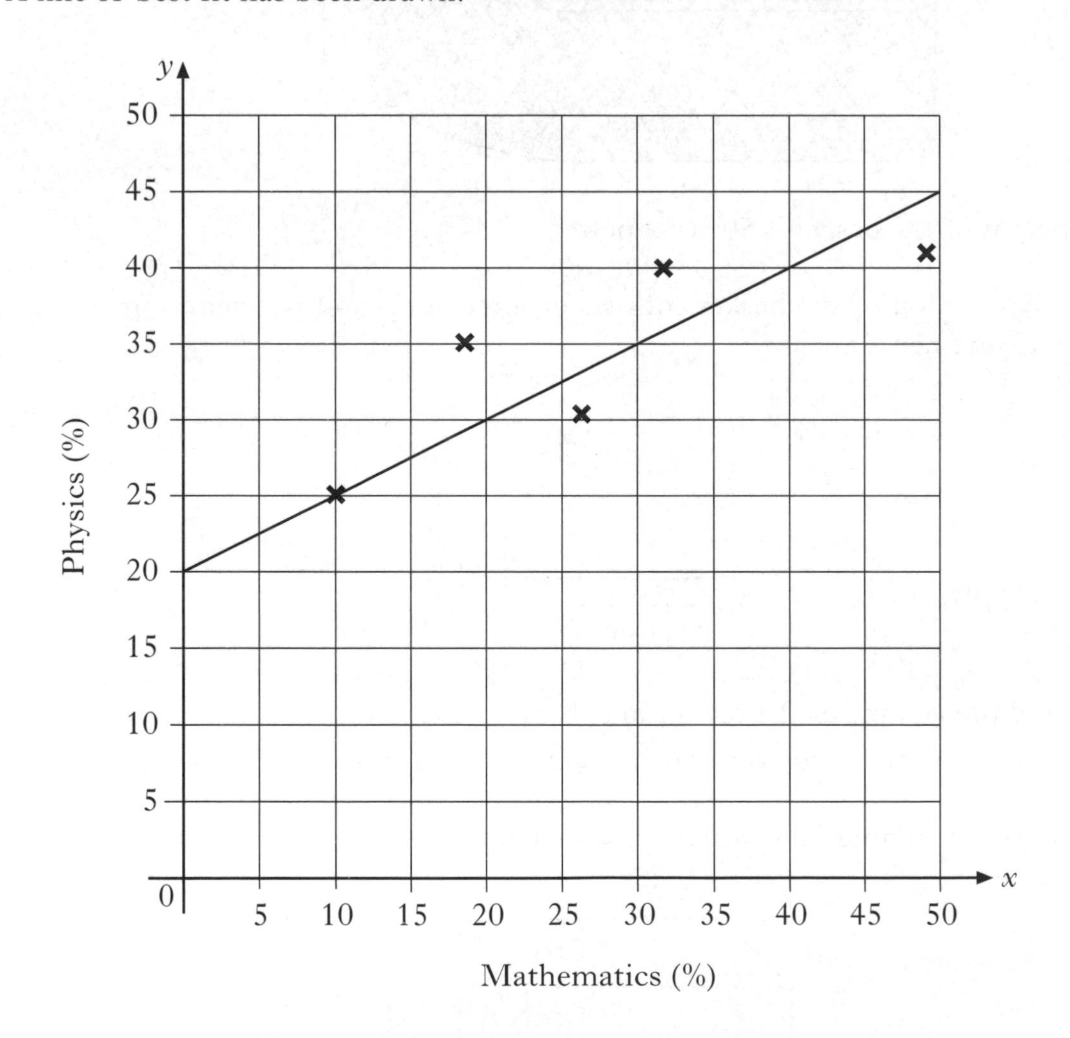

(*c*) Find the equation of the line of best fit. **3**

(*d*) Another pupil scored 76% in the mathematics test but was absent from the physics test.
Use your answer to part (*c*) to predict his physics mark. **1**

Marks

4. Suzie has a new mobile phone. She is charged x pence per minute for calls and y pence for each text she sends. During the first month her calls last a total of 280 minutes and she sends 70 texts. Her bill is £52·50.

(*a*) Write down an equation in x and y which satisfies the above condition. **1**

The next month she reduces her bill. She restricts her calls to 210 minutes and sends 40 texts. Her bill is £38·00.

(*b*) Write down a second equation in x and y which satisfies this condition. **1**

(*c*) Calculate the price per minute for a call and the price for each text sent. **4**

5. Triangle DEF is shown below.

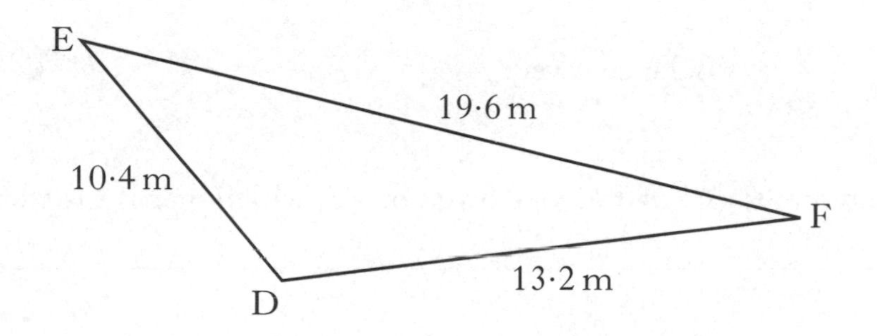

It has sides of length 10·4 metres, 13·2 metres and 19·6 metres.

Calculate the size of angle EDF.

Do not use a scale drawing. **3**

[Turn over

Marks

6. Below is a copy of part of David Leblanc's credit card statement.

Southern Star Credit

Name: David Leblanc — 12 May, 2008

Card Number: 4517 6767 2368 9001 — Credit Limit £3600

12 April 2008	Balance brought forward	£125·00
2 May 2008	Payment received	–50·00
	Balance	**A**
	Interest at 1·6%	**B**
5 May 2008	Bon Cave Wines	62·99
5 May 2008	Jacques Delicatessen	15·88
	Balance owed	**C**

Minimum payment: 3% of Balance owed or £5, whichever is greater.

(*a*) Calculate the amounts which would appear at **A**, **B** and **C**. **3**

(*b*) David makes the minimum payment.
How much does he pay? **2**

Marks

10. Irene works in the local chemist's shop.

One week she works 40 hours at her basic rate of pay and 3 hours overtime at double time.

Her gross pay for that week was £239·20.

Calculate Irene's basic hourly rate of pay. **3**

[*END OF QUESTION PAPER*]

[BLANK PAGE]

INTERMEDIATE 2

2009

[BLANK PAGE]

Marks

ALL questions should be attempted.

1. The number of goals scored one weekend by each team in the Football League is shown below.

0	1	1	2	1	0	0	5	0	1	3
0	2	2	1	1	3	0	0	2	4	1

(*a*) Construct a dotplot for the data. **2**

(*b*) The shape of the distribution is

A skewed to the right
B symmetric
C skewed to the left
D uniform.

Write down the letter that corresponds to the correct shape. **1**

2.

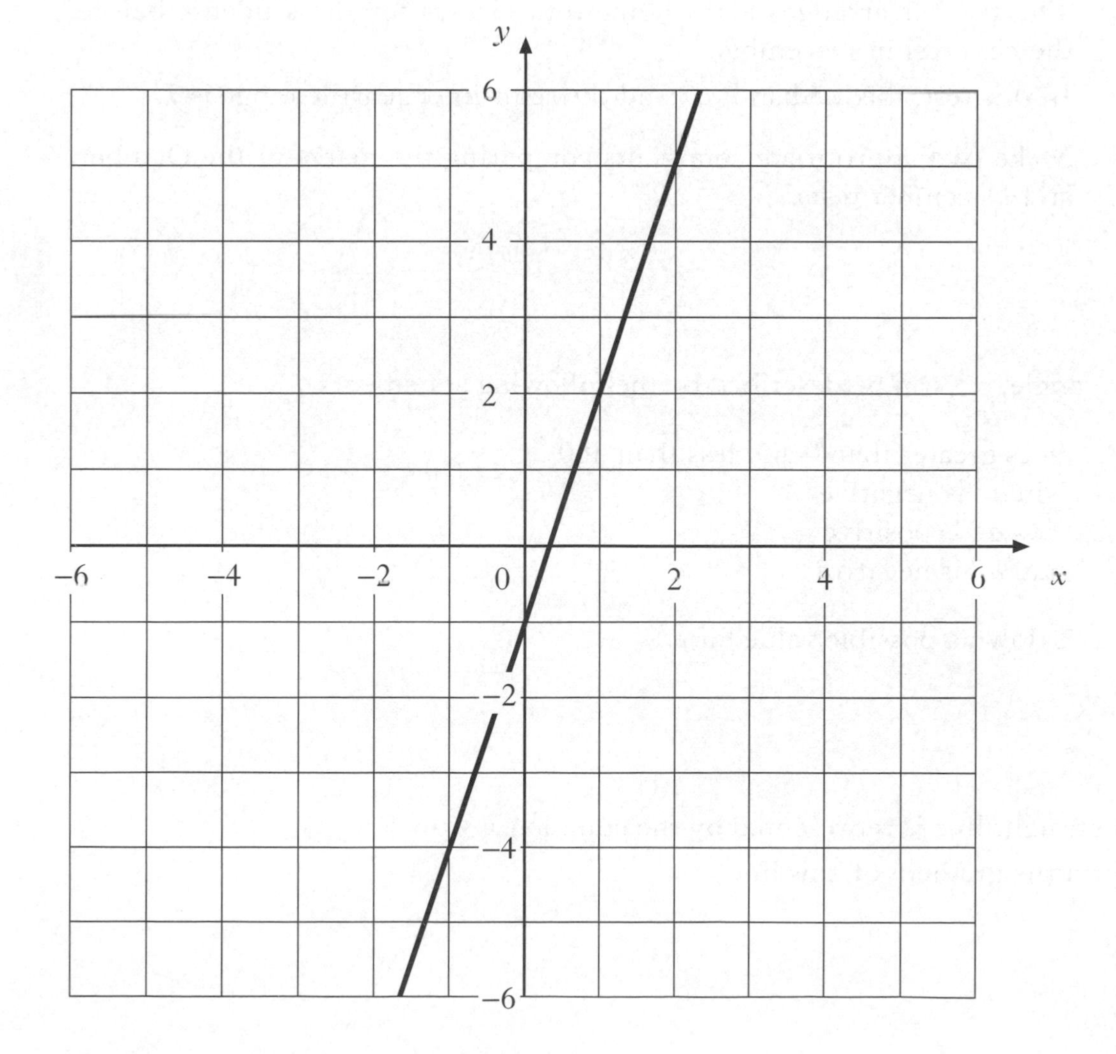

Find the equation of the straight line shown in the diagram. **3**

Marks

3. Factorise

$$x^2 - 5x - 24.$$ **2**

4. Multiply out the brackets and collect like terms.

$$(x + 5)(2x^2 - 3x - 1)$$ **3**

5. (*a*) The marks of a group of students in their October test are listed below.

41 56 68 59 43 37 70 58 61 47 75 66

Calculate:

(i) the median; **1**

(ii) the semi-interquartile range. **3**

(*b*) The teacher arranges extra homework classes for the students before the next test in December.

In this test, the median is 67 and the semi-interquartile range is 7.

Make **two** appropriate comments comparing the marks in the October and December tests. **2**

6. An angle, $a°$, can be described by the following statements.

- a is greater than 0 and less than 360
- $\sin a°$ is negative
- $\cos a°$ is positive
- $\tan a°$ is negative

Write down a possible value for a. **1**

7. A straight line is represented by the equation $x + y = 5$.

Find the gradient of this line. **2**

Marks

8. Five towns are represented by letters A, B, C, D and E in the tree diagram shown below.

The tree diagram represents routes between these five towns.

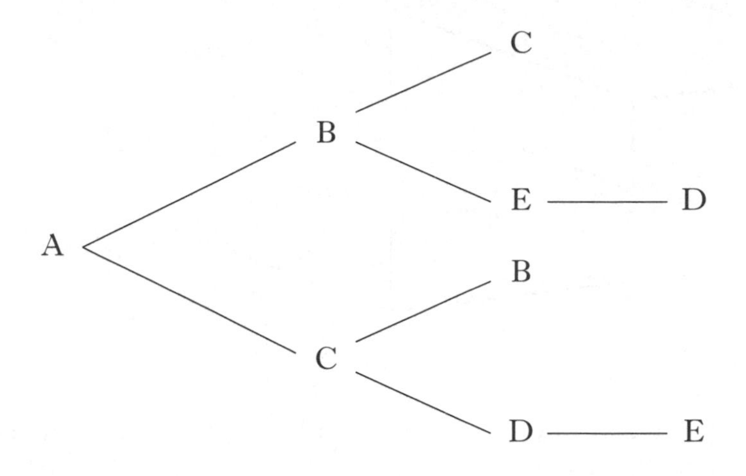

Draw a network diagram to represent the routes shown in the tree diagram. **2**

9. A company of window fitters uses a spreadsheet to show examples of how their prices are calculated.

	A	B	C	D	E
1	**Wendy's Window Fitters**			**Quotation for fitting windows**	
2	**VAT rate (%)**	17.5			
3					
4	**Window Size**	**Cost per window**	**Quantity of windows**	**Cost**	**Cost including VAT**
5					
6	30 cm by 30 cm	£50	3	£150	£176.25
7	50 cm by 70 cm	£75	4	£300	£352.50
8	120 cm by 100 cm	£120	2	£240	£282.00
9	90 cm by 150 cm	£125	1	£125	£146.88
10					
11					**£957.63**
12					

(*a*) Write down the formula used in cell E11. **1**

(*b*) The VAT rate in cell B2 is changed.
As a result, the values in column E are updated automatically.
Write down the formula used in cell E6. **2**

[Turn over for Question 10 on *Page six*

Marks

10. A three-dimensional solid is shown in the diagram below.

All dimensions are in centimetres.

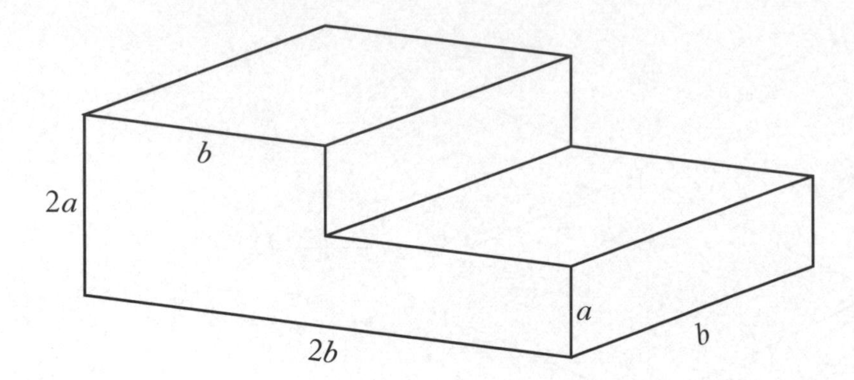

The surface area, S square centimetres, of this solid is given by the formula

$$S = 10ab + 4b^2.$$

(*a*) Calculate S when $a = 12$ and $b = 5$. **2**

(*b*) Calculate a when $S = 424$ and $b = 4$. **3**

[*END OF QUESTION PAPER*]

X101/204

NATIONAL QUALIFICATIONS 2009

THURSDAY, 21 MAY
2.05 PM – 3.35 PM

MATHEMATICS
INTERMEDIATE 2
Units 1, 2 and
Applications of Mathematics
Paper 2

Read carefully

1 **Calculators may be used in this paper.**
2 Full credit will be given only where the solution contains appropriate working.
3 Square-ruled paper is provided.

FORMULAE LIST

Sine rule: $\frac{a}{\sin A} = \frac{b}{\sin B} = \frac{c}{\sin C}$

Cosine rule: $a^2 = b^2 + c^2 - 2bc \cos A$ or $\cos A = \frac{b^2 + c^2 - a^2}{2bc}$

Area of a triangle: Area $= \frac{1}{2}ab \sin C$

Volume of a sphere: Volume $= \frac{4}{3}\pi r^3$

Volume of a cone: Volume $= \frac{1}{3}\pi r^2 h$

Volume of a cylinder: Volume $= \pi r^2 h$

Standard deviation: $s = \sqrt{\frac{\sum(x - \bar{x})^2}{n-1}} = \sqrt{\frac{\sum x^2 - (\sum x)^2/n}{n-1}}$, where n is the sample size.

ALL questions should be attempted.

Marks

1. A new book "Intermediate 2 Maths is Fun" was published in 2006.
There were 3000 sales of the book during that year.
Sales rose by 11% in 2007 then fell by 10% in 2008.

Were the sales in 2008 more or less than the sales in 2006?
You must give a reason for your answer. **3**

2. The heights, in centimetres, of seven netball players are given below.

173 176 168 166 170 180 171

For this sample, calculate:

(*a*) the mean; **1**

(*b*) the standard deviation. **3**

Show clearly all your working.

[Turn over

Marks

3. A company manufactures aluminium tubes.

The cross-section of one of the tubes is shown in the diagram below.

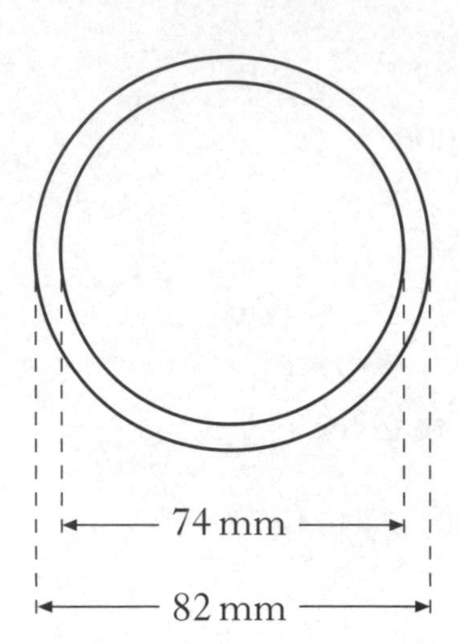

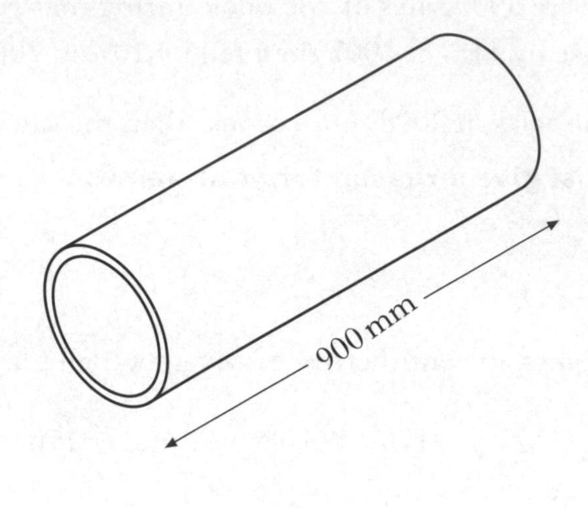

The inner diameter is 74 millimetres.

The outer diameter is 82 millimetres.

The tube is 900 millimetres long.

Calculate the volume of aluminium used to make the tube.

Give your answer correct to three significant figures. 5

4. There are 14 cars and 60 passengers on the morning crossing of the ferry from Wemyss Bay to Rothesay. The total takings are £344·30.

(*a*) Let x pounds be the cost for a car and y pounds be the cost for a passenger.

Write down an equation in x and y which satisfies the above condition. 1

(*b*) There are 21 cars and 40 passengers on the evening crossing of the ferry. The total takings are £368·95.

Write down a second equation in x and y which satisfies this condition. 1

(*c*) Find the cost for a car and the cost for a passenger on the ferry. 4

Marks

5. A pet shop manufactures protective dog collars.

In the diagram below the shaded area represents one of these collars.

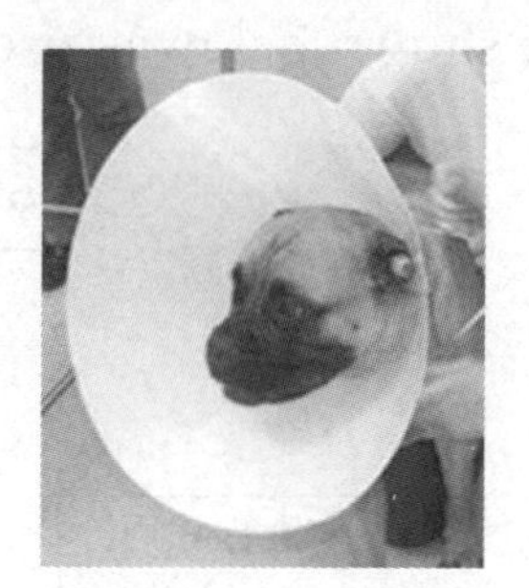

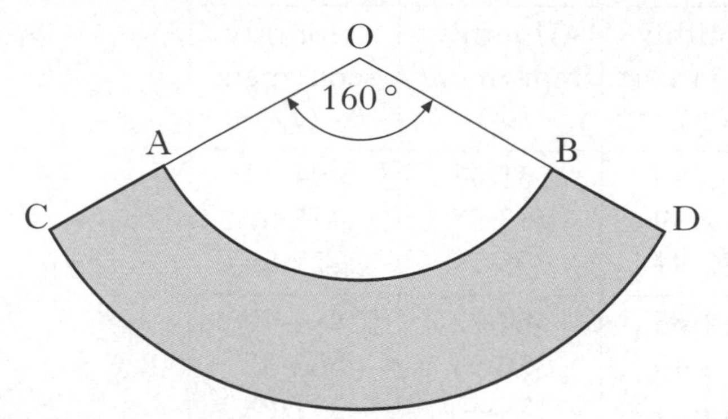

AB and CD are arcs of the circles with centres at O.

The radius, OA, is 10 inches and the radius, OC, is 18 inches.

Angle AOB is 160°.

Calculate the area of a collar. **4**

6. The Bermuda triangle is an area in the Atlantic Ocean where many planes and ships have mysteriously disappeared.

Its vertices are at Bermuda (B), Miami (M) and Puerto Rico (P).

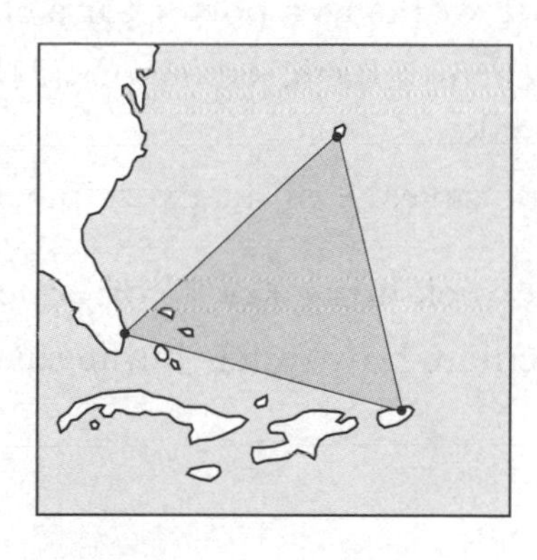

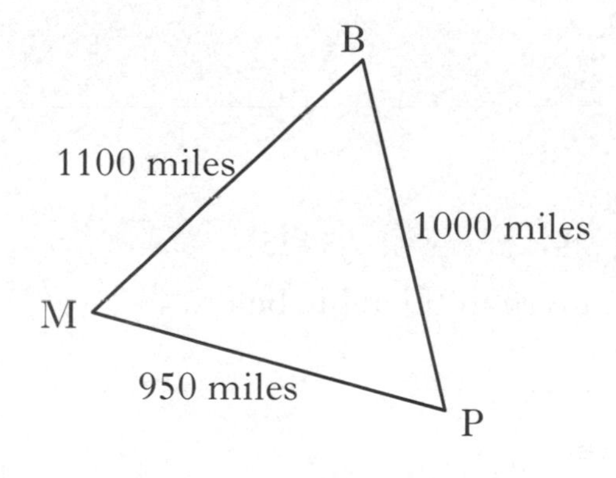

Calculate the size of angle BPM. **3**

[Turn over

Marks

7. The table shown below is used to calculate loan repayments.

		60 months	48 months	24 months
		Monthly repayment (£)	Monthly repayment (£)	Monthly repayment (£)
With payment protection	£20 000	463·85	551·43	994·23
	£15 000	347·89	413·57	745·67
	£7500	173·94	206·79	372·84
Without payment protection	£20 000	384·65	467·72	884·47
	£15 000	288·49	350·79	663·35
	£7500	144·24	175·40	331·68

Samir wishes to borrow £15 000.

How much will the loan cost him if he repays it over 48 months, with payment protection? **3**

8. Jamie works as a potter for a company which makes china ornaments.

He earns a basic salary of £218 per week plus 80 pence for every ornament he makes.

Jamie saves $\frac{2}{5}$ of his gross pay every week.

One week he makes 40 ornaments.

Calculate how much Jamie saves that week. **3**

9. Anna earns £42 000 per year. She has tax allowances of £5425.

The rates of tax applicable for the year are given in the table below.

Taxable income (£)	**Rate**
On the first £34 600	20%
On any income over £34 600	40%

How much is Anna's **monthly** tax bill? **5**

Marks

10. For reasons of safety, a building is supported by two wooden struts, represented by DB and DC in the diagram below.

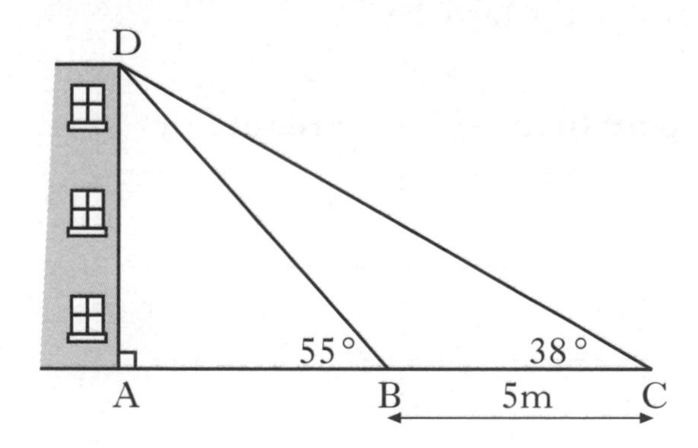

Angle ABD = 55°.

Angle BCD = 38°.

BC is 5 metres.

Calculate the height of the building represented by AD. **5**

11. A railway goes through an underground tunnel.

The diagram below shows the cross-section of the tunnel. It consists of part of a circle with a horizontal base.

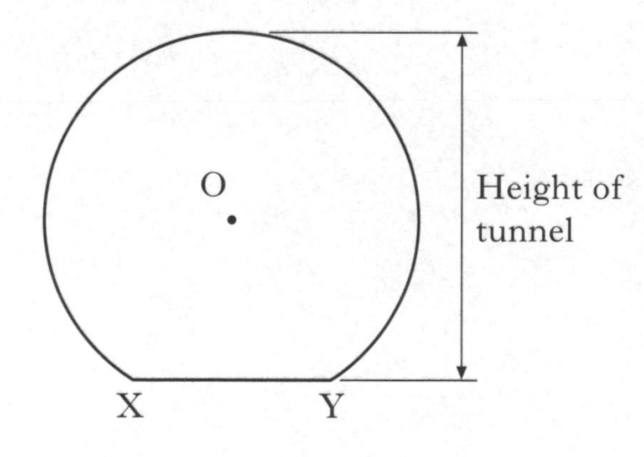

- The centre of the circle is O.
- XY is a chord of the circle.
- XY is 1·8 metres.
- The radius of the circle is 1·7 metres.

Find the height of the tunnel. **4**

[Turn over for Question 12 on *Page eight*

Marks

12. The amount of money spent by each pupil in a school tuck shop is recorded.

The data collected is shown in the table below.

Amount spent (pence)	**Frequency**
1– 50	42
51–100	64
101–150	35
151–200	18
201–250	12
251–300	10

Calculate the mean amount of money spent by pupils in the tuck shop. **5**

[*END OF QUESTION PAPER*]

INTERMEDIATE 2

2010

[BLANK PAGE]

X101/202

NATIONAL QUALIFICATIONS 2010

FRIDAY, 21 MAY
1.00 PM – 1.45 PM

MATHEMATICS
INTERMEDIATE 2
Units 1, 2 and
Applications of Mathematics
Paper 1
(Non-calculator)

Read carefully

1 **You may NOT use a calculator.**

2 Full credit will be given only where the solution contains appropriate working.

3 Square-ruled paper is provided.

FORMULAE LIST

Sine rule: $\frac{a}{\sin A} = \frac{b}{\sin B} = \frac{c}{\sin C}$

Cosine rule: $a^2 = b^2 + c^2 - 2bc \cos A$ or $\cos A = \frac{b^2 + c^2 - a^2}{2bc}$

Area of a triangle: Area $= \frac{1}{2} ab \sin C$

Volume of a sphere: Volume $= \frac{4}{3} \pi r^3$

Volume of a cone: Volume $= \frac{1}{3} \pi r^2 h$

Volume of a cylinder: Volume $= \pi r^2 h$

Standard deviation: $s = \sqrt{\frac{\sum (x - \bar{x})^2}{n-1}} = \sqrt{\frac{\sum x^2 - (\sum x)^2 / n}{n-1}}$, where n is the sample size.

Marks

ALL questions should be attempted.

1.

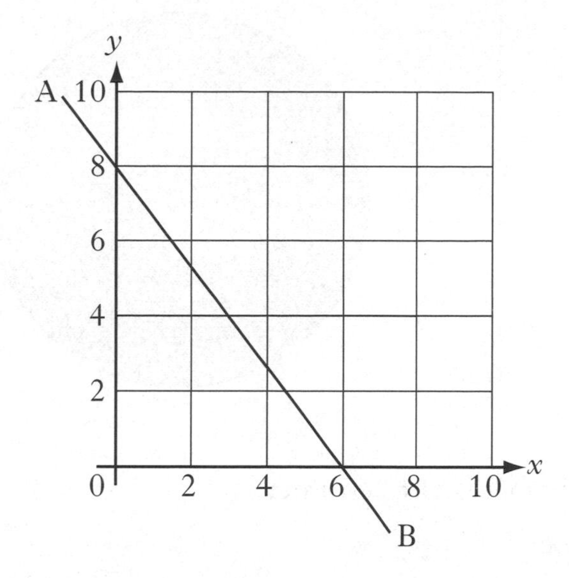

Find the equation of the straight line AB shown in the diagram. **3**

2. The pupils in a primary class record their shoe sizes as shown below.

8	7	6	5	6
5	7	11	7	7
7	8	7	9	6
8	6	5	9	7

(*a*) Construct a frequency table from the above data and add a cumulative frequency column. **2**

(*b*) For this data, find:

(i) the median; **1**

(ii) the lower quartile; **1**

(iii) the upper quartile. **1**

(*c*) Construct a boxplot for this data. **2**

[Turn over

Marks

3. The diagram below represents a sphere.

The sphere has a diameter of 6 centimetres.

Calculate its volume.

Take $\pi = 3\cdot14$. **2**

4. (*a*) Factorise

$$x^2 + x - 6.$$ **2**

(*b*) Multiply out the brackets and collect like terms.

$$(3x + 2)(x^2 + 5x - 1)$$ **3**

Marks

5. The diagram shows a network of streets connecting certain landmarks in a town centre.

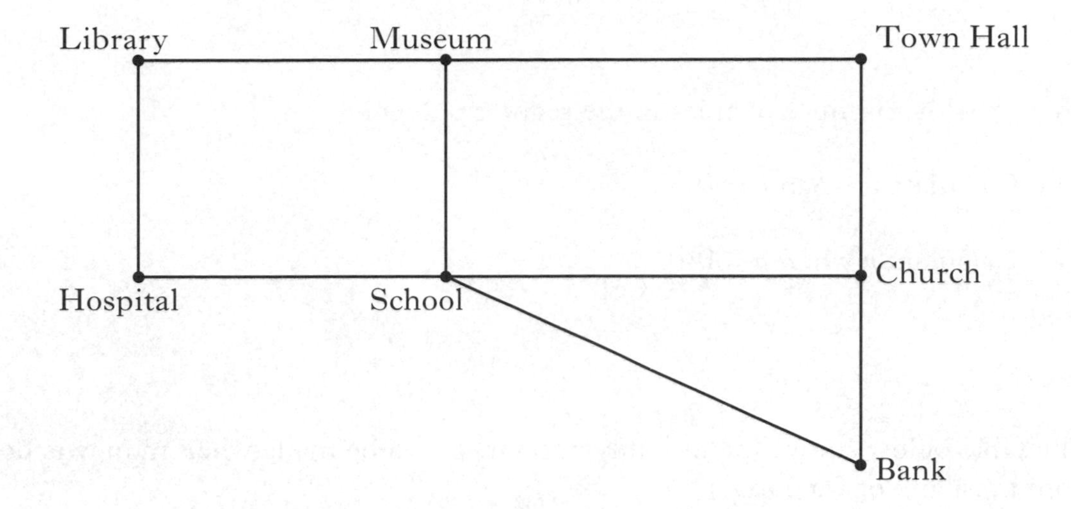

A bin lorry has to collect rubbish along every street shown.

Is it possible to do this without travelling any street more than once? Explain your answer. **2**

6.

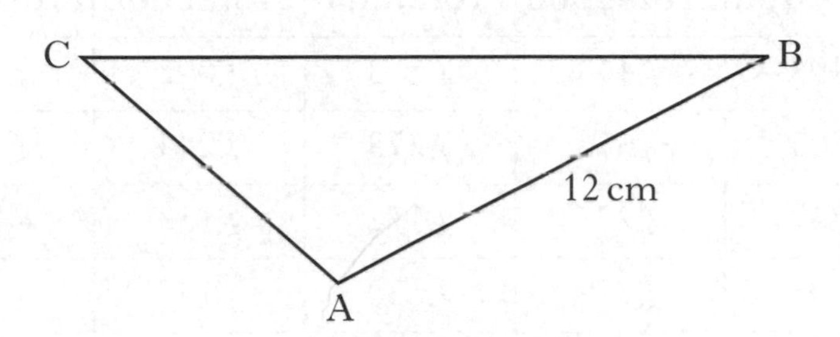

In triangle ABC, AB = 12 centimetres, $\sin C = \frac{1}{2}$ and $\sin B = \frac{1}{3}$.

Find the length of side AC. **3**

[Turn over for Questions 7 and 8 on *Page six*

Marks

7. The size of each angle, $a°$, in a regular polygon is given by the formula

$$a = 180 - \frac{360}{n},$$

where n is the number of sides in the regular polygon.

(*a*) Calculate a when $n = 10$. **2**

(*b*) Calculate n when $a = 140$. **3**

8. The table below shows the monthly repayments to be made when money is borrowed from the Bank of Caledonia.

Repayments can be made with or without loan protection.

	Monthly repayments: Bank of Caledonia					
	24 months		**36 months**		**48 months**	
Loan Amount	**With Loan Protection**	**Without Loan Protection**	**With Loan Protection**	**Without Loan Protection**	**With Loan Protection**	**Without Loan Protection**
£10 000	£495	£445	£343	£305	£277	£237
£8000	£395	£356	£275	£244	£222	£190
£5000	£247	£223	£172	£153	£139	£119
£4000	£198	£179	£138	£123	£111	£95

Jeremy borrows £8000 over 36 months **without** loan protection.

After 28 months, he is made redundant and is unable to pay the remainder of the loan.

His brother, Peter, agrees to make the remaining payments.

How much does Peter pay in total? **3**

[*END OF QUESTION PAPER*]

X101/204

NATIONAL QUALIFICATIONS 2010

FRIDAY, 21 MAY
2.05 PM – 3.35 AM

MATHEMATICS
INTERMEDIATE 2
Units 1, 2 and
Applications of Mathematics
Paper 2

Read carefully

1 **Calculators may be used in this paper.**

2 Full credit will be given only where the solution contains appropriate working.

3 Square-ruled paper is provided.

LI X101/204 6/8610

FORMULAE LIST

Sine rule: $\frac{a}{\sin A} = \frac{b}{\sin B} = \frac{c}{\sin C}$

Cosine rule: $a^2 = b^2 + c^2 - 2bc \cos A$ or $\cos A = \frac{b^2 + c^2 - a^2}{2bc}$

Area of a triangle: Area $= \frac{1}{2}ab \sin C$

Volume of a sphere: Volume $= \frac{4}{3}\pi r^3$

Volume of a cone: Volume $= \frac{1}{3}\pi r^2 h$

Volume of a cylinder: Volume $= \pi r^2 h$

Standard deviation: $s = \sqrt{\frac{\Sigma(x - \bar{x})^2}{n-1}} = \sqrt{\frac{\Sigma x^2 - (\Sigma x)^2 / n}{n-1}}$, where n is the sample size.

Marks

ALL questions should be attempted.

1. An industrial machine costs £176 500.

Its value depreciates by 4·25% each year.

How much is it worth after 3 years?

Give your answer correct to **three** significant figures. **4**

2. Paul conducts a survey to find the most popular school lunch.

- 30 pupils vote for Pasta
- 40 pupils vote for Baked Potato
- 2 pupils vote for Salad

Paul wishes to draw a pie chart to illustrate his data. How many degrees must he use for each sector in his pie chart?

Do not draw the pie chart. **2**

3. The scattergraph shows the taxi fare, p pounds, plotted against the distance travelled, m miles. A line of best fit has been drawn.

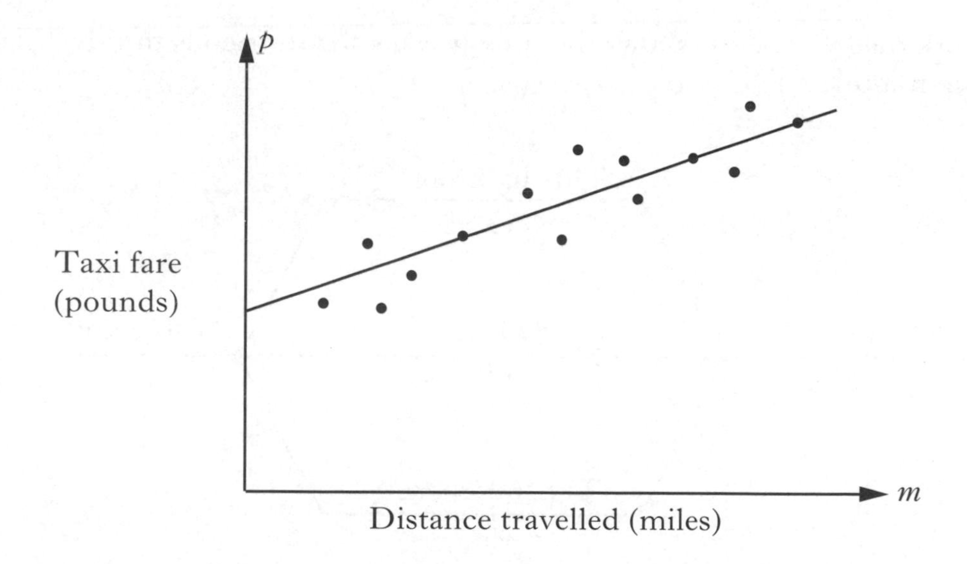

The equation of the line of best fit is $p = 2 + 1{\cdot}5\,m$.

Use this equation to predict the taxi fare for a journey of 6 miles. **1**

[Turn over

Marks

4. A rugby team scored the following points in a series of matches.

13 7 0 9 7 8 5

(*a*) For this sample, calculate:

(i) the mean; **1**

(ii) the standard deviation. **3**

Show clearly all your working.

The following season, the team appoints a new coach.

A similar series of matches produces a mean of 27 and a standard deviation of 3·25.

(*b*) Make two valid comparisons about the performance of the team under the new coach. **2**

5. Solve algebraically the system of equations

$$2x - 5y = 24$$
$$7x + 8y = 33.$$

3

6. The network diagram below shows the time it takes **three** friends to tidy a flat. All times are in minutes.

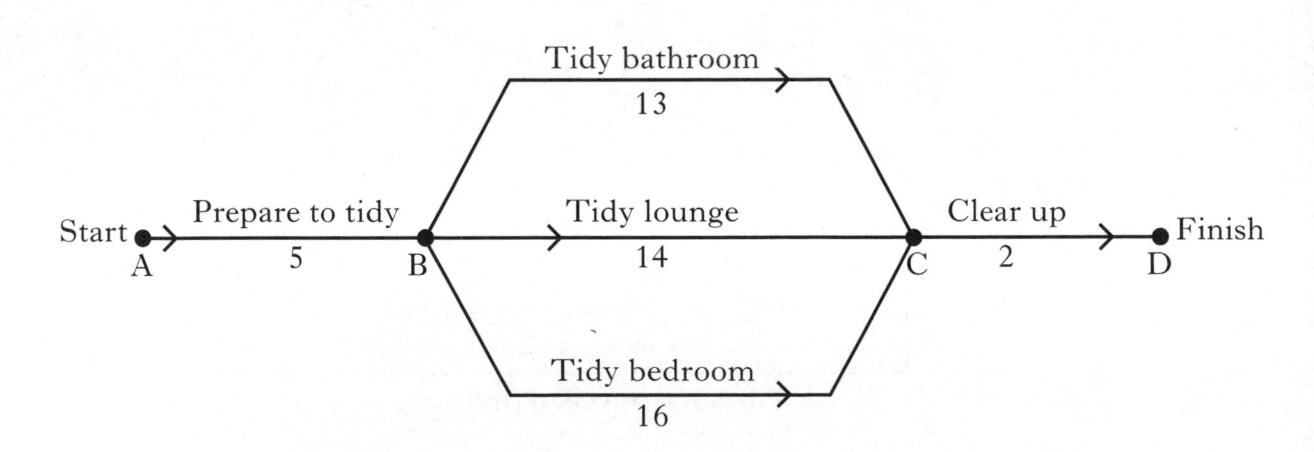

Guests are due to arrive in 20 minutes.

Will the flat be tidy on time?

Give a reason for your answer. **1**

Marks

7. Sam sells used cars. She keeps a record of her profits on a spreadsheet.

	A	B	C	D	E
1	**Make of Car**	**Cost Price**	**Selling Price**	**Profit**	**Profit (%)**
2					
3	**Sultan**	£2500	£3800	£1300	52
4	**Astral 4**	£3600	£4800		
5	**Ventra**	£2000	£3000		
6	**Satellite 5**	£7250	£8120		
7	**Phoenix**	£2800	£3080		

(*a*) What formula would be used to enter the profit in cell D6? **1**

(*b*) The result of the formula =D6/B6*100 is to appear in cell E6.

What value will appear in cell E6? **3**

8. The cost of electricity per quarter to a sample of homes in Bellrock Avenue was recorded.

The results are shown in the histogram below.

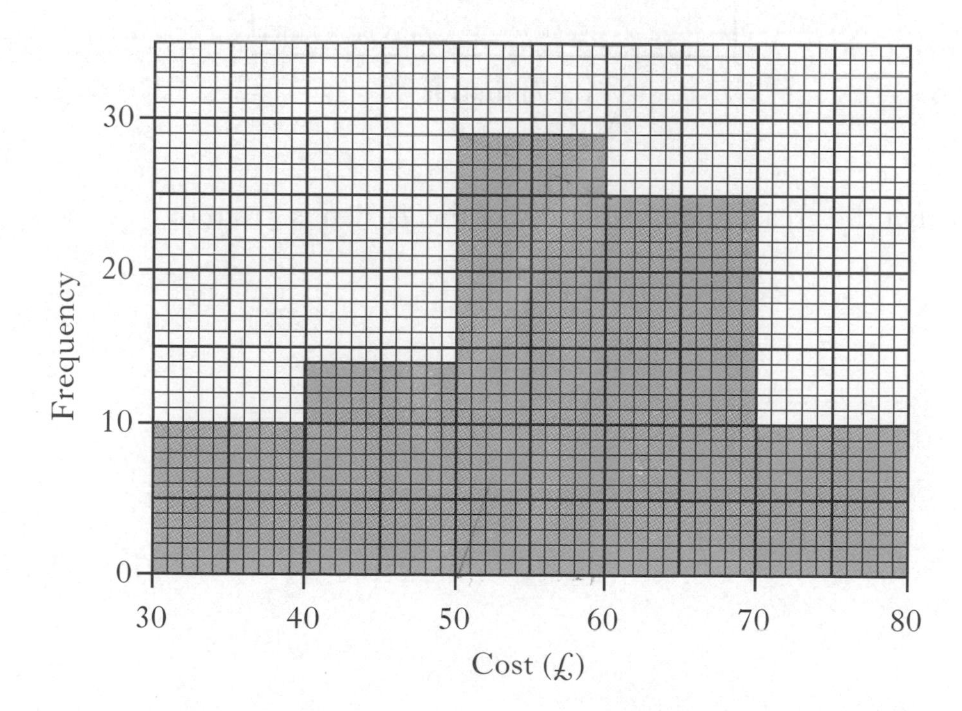

Estimate the value of the mode. **1**

[Turn over

Marks

9. The ends of a magazine rack are identical.

Each end is a sector of a circle with radius 14 centimetres.
The angle in each sector is 65 °.

The sectors are joined by two rectangles, each with length 40 centimetres.

The exterior is covered by material.
What area of material is required?

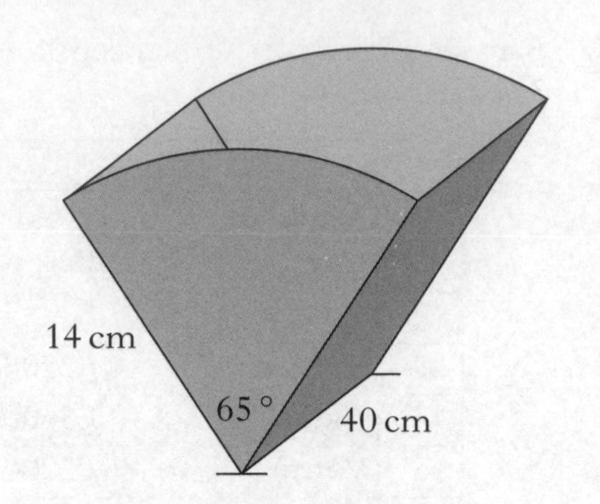

4

10. The diagram below represents a rectangular garden with length $(x + 7)$ metres and breadth $(x + 3)$ metres.

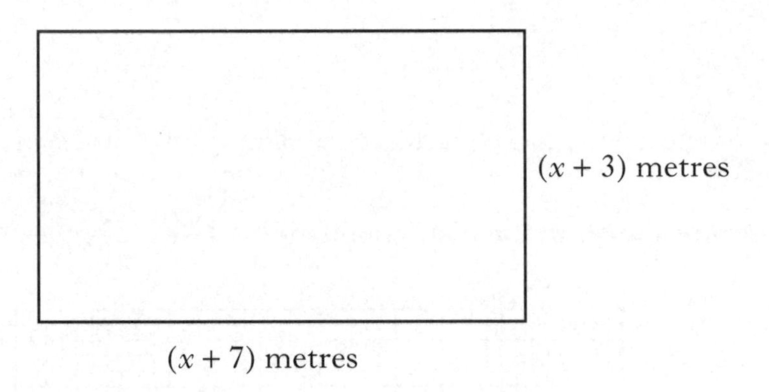

Show that the area, A square metres, of the garden is given by

$$A = x^2 + 10x + 21.$$

2

Marks

11. A cylindrical container has a volume of 3260 cubic centimetres.

The radius of the cross section is 6·4 centimetres.

Calculate the height of the cylinder. **3**

12. Two ships have located a wreck on the sea bed.

In the diagram below, the points P and Q represent the two ships and the point R represents the wreck.

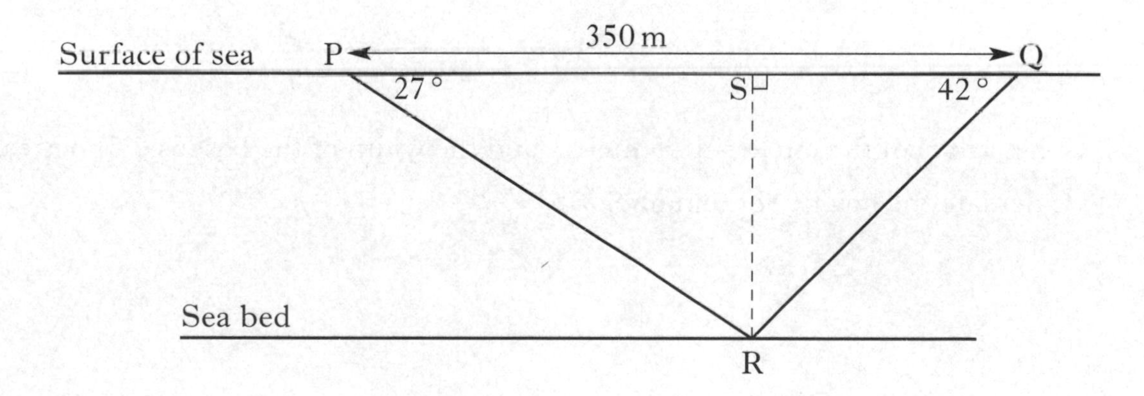

The angle of depression of R from P is 27°.
The angle of depression of R from Q is 42°.
The distance PQ is 350 metres.

Calculate QS, the distance ship Q has to travel to be directly above the wreck.

Do not use a scale drawing. **5**

[Turn over

Marks

13. Ocean World has an underwater viewing tunnel.

The diagram below shows the cross-section of the tunnel. It consists of part of a circle with a horizontal base.

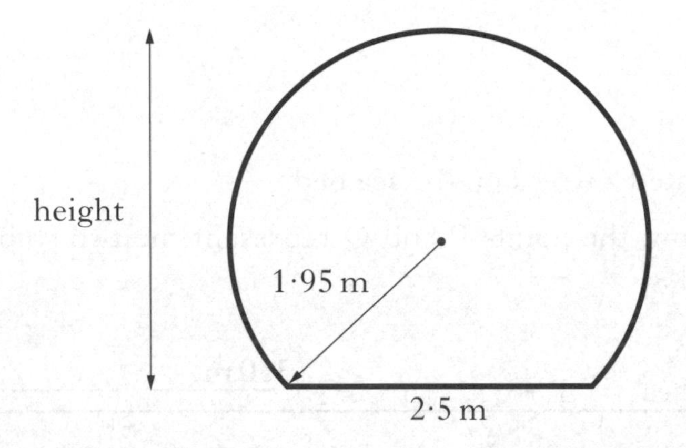

The radius of the circle is 1·95 metres and the width of the base is 2·5 metres.

Calculate the height of the tunnel. **4**

Marks

14. Shaheen works in a call centre. Her basic rate of pay is £6·40 per hour.

She is paid time and a half for working overtime in the evening and double time for working overtime at the weekend.

One week she works 35 hours at the basic rate and 6 hours overtime in the evening. She also works overtime at the weekend.

Shaheen's gross pay for the week is £320.

How many hours does she work at the weekend? **4**

15. The marks of a group of students in their Intermediate 2 Mathematics examination were recorded.

The cumulative frequency results are shown below.

Marks (m)	*Cumulative Frequency*
$0 \leq m < 10$	1
$10 \leq m < 20$	5
$20 \leq m < 30$	12
$30 \leq m < 40$	30
$40 \leq m < 50$	48
$50 \leq m < 60$	55
$60 \leq m < 70$	59
$70 \leq m < 80$	60

(*a*) Using squared paper, draw a cumulative frequency diagram for this data. **3**

(*b*) From your diagram, estimate:

(i) the lower quartile; **1**

(ii) the upper quartile. **1**

(*c*) Calculate the semi-interquartile range. **1**

[*END OF QUESTION PAPER*]

[BLANK PAGE]

INTERMEDIATE 2

2011

[BLANK PAGE]

X101/202

NATIONAL QUALIFICATIONS 2011

WEDNESDAY, 18 MAY
1.00 PM – 1.45 PM

MATHEMATICS
INTERMEDIATE 2
Units 1, 2 and
Applications of Mathematics
Paper 1
(Non-calculator)

Read carefully

1 **You may NOT use a calculator.**

2 Full credit will be given only where the solution contains appropriate working.

3 Square-ruled paper is provided. If you make use of this, you should write your name on it clearly and put it inside your answer booklet.

LI X101/202 6/9010

FORMULAE LIST

Sine rule: $\frac{a}{\sin A} = \frac{b}{\sin B} = \frac{c}{\sin C}$

Cosine rule: $a^2 = b^2 + c^2 - 2bc \cos A$ or $\cos A = \frac{b^2 + c^2 - a^2}{2bc}$

Area of a triangle: $\text{Area} = \frac{1}{2}ab \sin C$

Volume of a sphere: $\text{Volume} = \frac{4}{3}\pi r^3$

Volume of a cone: $\text{Volume} = \frac{1}{3}\pi r^2 h$

Volume of a cylinder: $\text{Volume} = \pi r^2 h$

Standard deviation: $s = \sqrt{\frac{\sum(x - \bar{x})^2}{n-1}} = \sqrt{\frac{\sum x^2 - (\sum x)^2 / n}{n-1}}$, where n is the sample size.

Marks

ALL questions should be attempted.

1. Sandi takes the bus to work each day.

Over a two week period, she records the number of minutes the bus is late each day. The results are shown below.

5 6 15 0 6 11 2 9 8 7

(*a*) From the above data, find:

(i) the median; **1**

(ii) the lower quartile; **1**

(iii) the upper quartile. **1**

(*b*) Construct a boxplot for the data. **2**

Sandi decides to take the train over the next two week period and records the number of minutes the train is late each day.

The boxplot, drawn below, was constructed for the new data.

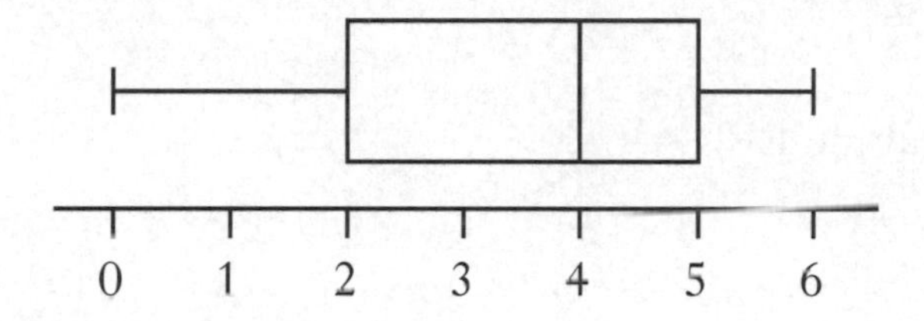

(*c*) Compare the two boxplots and comment. **1**

2. Factorise

$$x^2 - 4x - 21.$$ **2**

3. Multiply out the brackets and collect like terms.

$$5x + (3x + 2)(2x - 7)$$ **3**

[Turn over

Marks

4. A circle, centre O, is shown below.

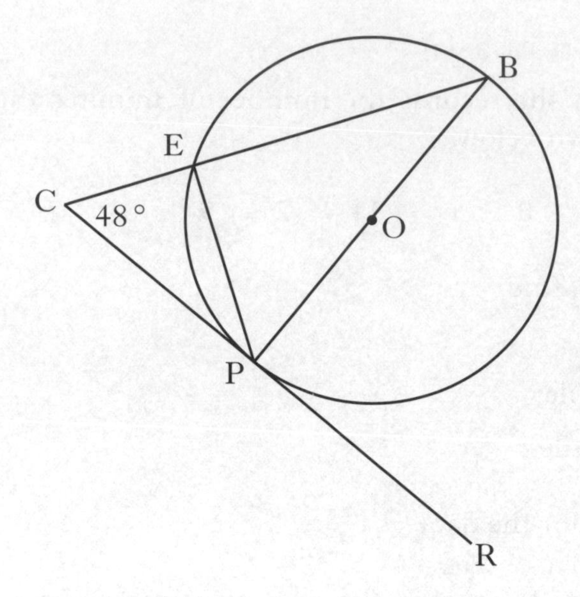

In the circle

- PB is a diameter
- CR is a tangent to the circle at point P
- Angle BCP is 48 °.

Calculate the size of angle EPR. **3**

5. The approximate stopping distance of a car can be found by using the formula

$$D = \frac{1}{3}\left(S + \frac{S^2}{20}\right)$$

where D metres is the approximate stopping distance
and S miles per hour is the speed before braking.

Calculate the approximate stopping distance when the speed before braking is 30 miles per hour. **3**

Marks

6. Below is the summary part of Geetha's Credit Card statement at the end of May.

Briggs Bank

CREDIT CARD STATEMENT

Summary as at 21 May 2011

Credit Limit	£4000	
Available Credit	£3760	
Balance from previous statement		£0·00
New Transactions		£240·00
Interest		£0·00
Balance owed		£240·00
Minimum payment due		£7·20
Payment due date		15 June 2011

Interest will be charged at 1% per month on any outstanding balance.

Geetha pays the minimum payment.

She does not use the credit card again.

What is the "Balance owed" in her next statement? **2**

[Turn over

Marks

7.

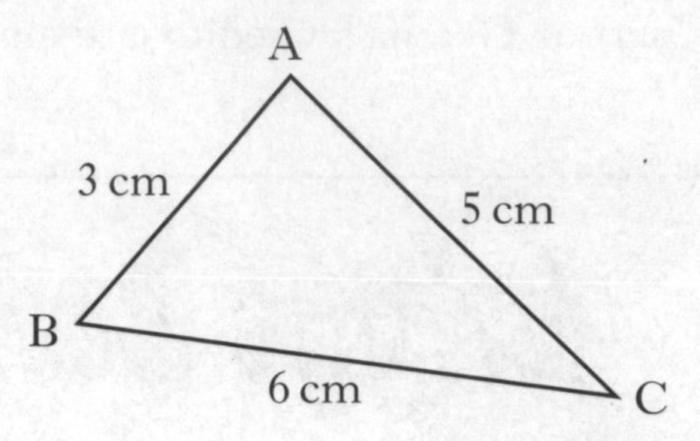

In triangle ABC, show that $\cos B = \frac{5}{9}$. **3**

8. A straight line is represented by the equation $y = mx + c$.

Sketch a possible straight line graph to illustrate this equation when $m > 0$ and $c < 0$. **2**

Marks

9. A catering company supplies the airports at Aberdeen (A), Edinburgh (E), Glasgow (G), Newcastle (N) and Prestwick (P). The network diagram below represents the distances in miles by road between the airports.

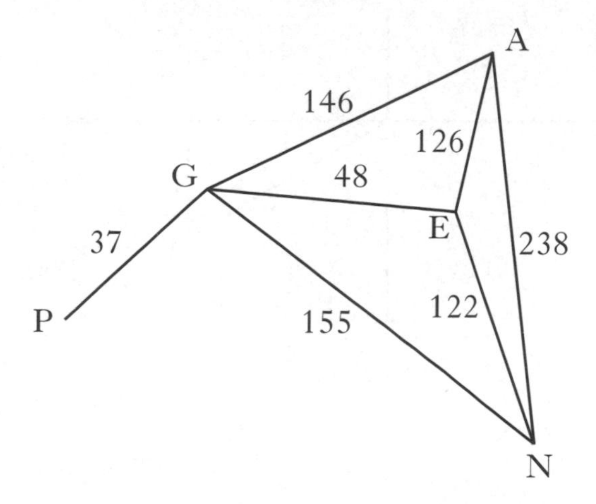

Catering supplies are distributed by van from Prestwick to the other airports. The van does not need to return to Prestwick.

(*a*) Copy and complete the tree diagram to show **all** the possible routes the van can take.

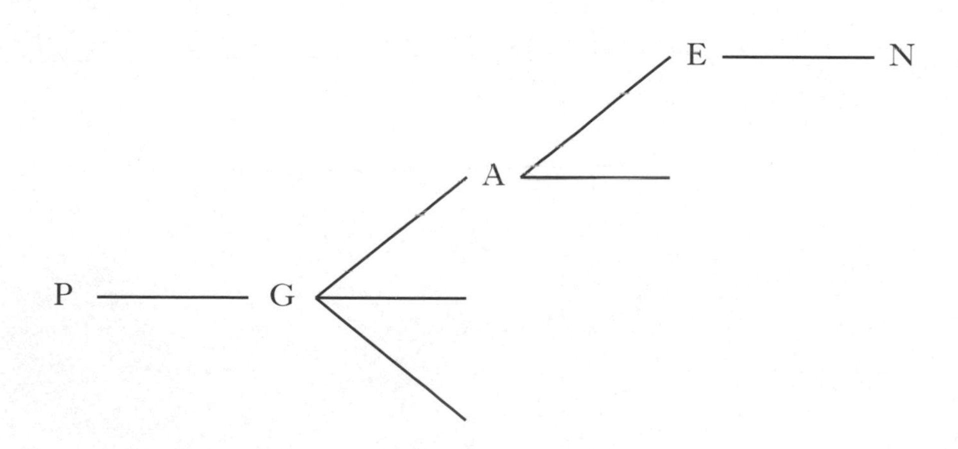

3

(*b*) The van driver decides he wants to finish the journey at **Newcastle**.

What is the shortest distance he has to drive to finish the journey at Newcastle? Explain your answer.

2

[Turn over for Question 10 on *Page eight*

Marks

10.

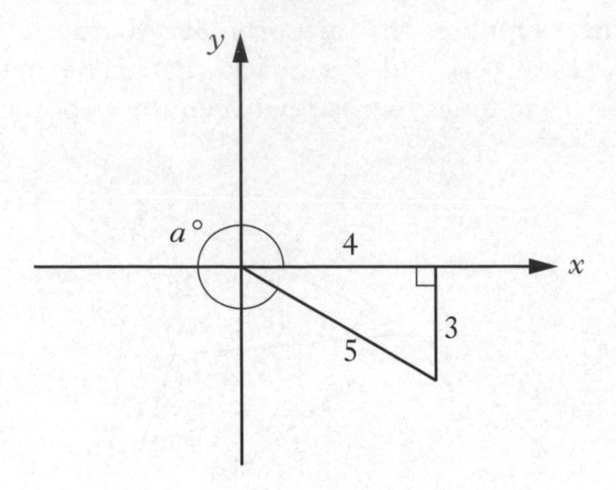

Write down the value of $\cos a°$. **1**

[*END OF QUESTION PAPER*]

X101/204

NATIONAL QUALIFICATIONS 2011

WEDNESDAY, 18 MAY
2.05 PM – 3.35 PM

MATHEMATICS
INTERMEDIATE 2
Units 1, 2 and
Applications of Mathematics
Paper 2

Read carefully

1 **Calculators may be used in this paper.**

2 Full credit will be given only where the solution contains appropriate working.

3 Square-ruled paper is provided. If you make use of this, you should write your name on it clearly and put it inside your answer booklet.

FORMULAE LIST

Sine rule: $\frac{a}{\sin A} = \frac{b}{\sin B} = \frac{c}{\sin C}$

Cosine rule: $a^2 = b^2 + c^2 - 2bc \cos A$ or $\cos A = \frac{b^2 + c^2 - a^2}{2bc}$

Area of a triangle: Area $= \frac{1}{2}ab \sin C$

Volume of a sphere: Volume $= \frac{4}{3}\pi r^3$

Volume of a cone: Volume $= \frac{1}{3}\pi r^2 h$

Volume of a cylinder: Volume $= \pi r^2 h$

Standard deviation: $s = \sqrt{\frac{\sum(x-\overline{x})^2}{n-1}} = \sqrt{\frac{\sum x^2 - (\sum x)^2/n}{n-1}}$, where n is the sample size.

Marks

ALL questions should be attempted.

1.

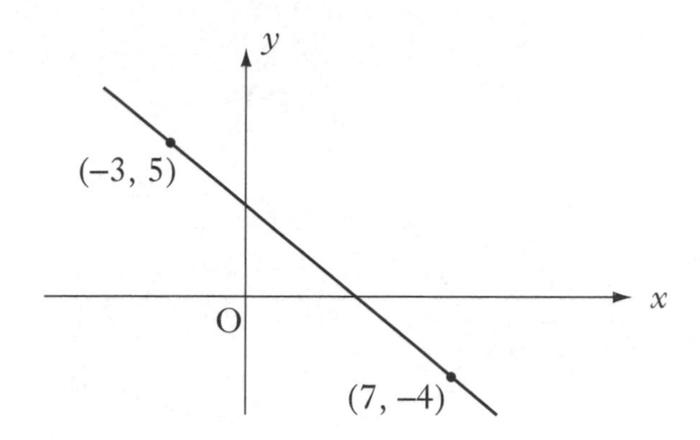

Calculate the gradient of the straight line passing through the points (–3, 5) and (7, –4). **1**

2. It is estimated that house prices will increase at the rate of 3·15% per annum.

A house is valued at £134 750. If its value increases at the predicted rate, calculate its value after 3 years.

Give your answer correct to **four** significant figures. **4**

3. The Battle of Largs in 1263 is commemorated by a monument known as The Pencil.

This monument is in the shape of a cylinder with a cone on top.

The cylinder part has diameter 3 metres and height 15 metres.

(*a*) Calculate the volume of the **cylinder** part of The Pencil. **2**

The volume of the **cone** part of The Pencil is 5·7 cubic metres.

(*b*) Calculate the **total** height of The Pencil. **3**

[Turn over

Marks

4. The diagram below shows a sector of a circle, centre C.

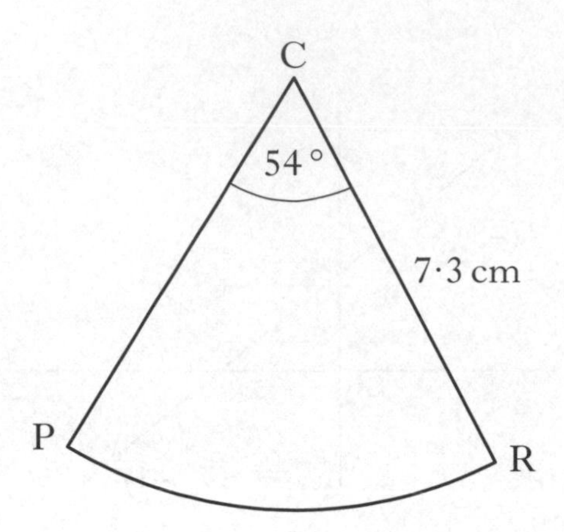

The radius of the circle is 7·3 centimetres and angle PCR is 54 °.

Calculate the area of the sector PCR. **3**

5. A sample of six boxes contains the following numbers of pins per box.

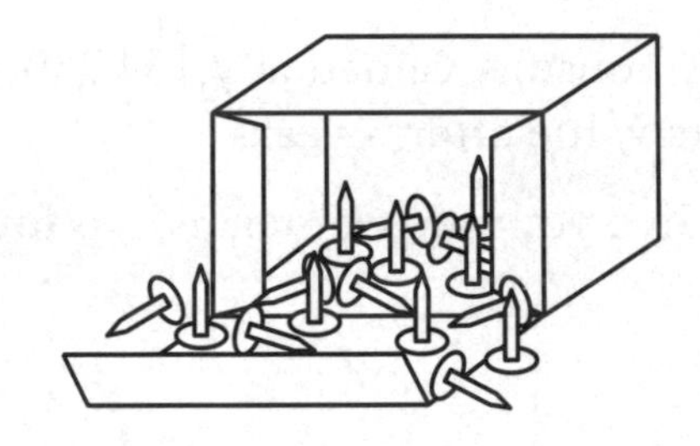

43 39 41 40 39 44

(*a*) For the above data, calculate:

(i) the mean; **1**

(ii) the standard deviation. **3**

The company which produces the pins claims that "the mean number of pins per box is 40 ± 2 and the standard deviation is less than 3".

(*b*) Does the data in part (*a*) support the claim made by the company?

Give reasons for your answer. **2**

Marks

6. Alan is taking part in a quiz. He is awarded x points for each correct answer and y points for each wrong answer. During the quiz, Alan gets 24 questions correct and 6 wrong. He scores 60 points.

(*a*) Write down an equation in x and y which satisfies the above condition. **1**

Helen also takes part in the quiz. She gets 20 questions correct and 10 wrong. She scores 40 points.

(*b*) Write down a second equation in x and y which satisfies this condition. **1**

(*c*) Calculate the score for David who gets 17 correct and 13 wrong. **4**

7. The table below gives the **monthly** repayments from three different banks on a £10 000 loan repaid over **five years**.

Name of Bank	Monthly Repayments	
	With payment protection	**Without payment protection**
Savewell	£245·39	£214·39
Finesave	£260·58	£205·65
Wisespend	£263·17	£214·70

Emily borrowed £10 000 and paid it back over five years. The cost of the loan was £2339. Which bank was the loan from and did she take it with or without payment protection? **3**

[Turn over

Marks

8. In a race, organisers record how long each runner takes to complete the course. The results are shown in the cumulative frequency curve below.

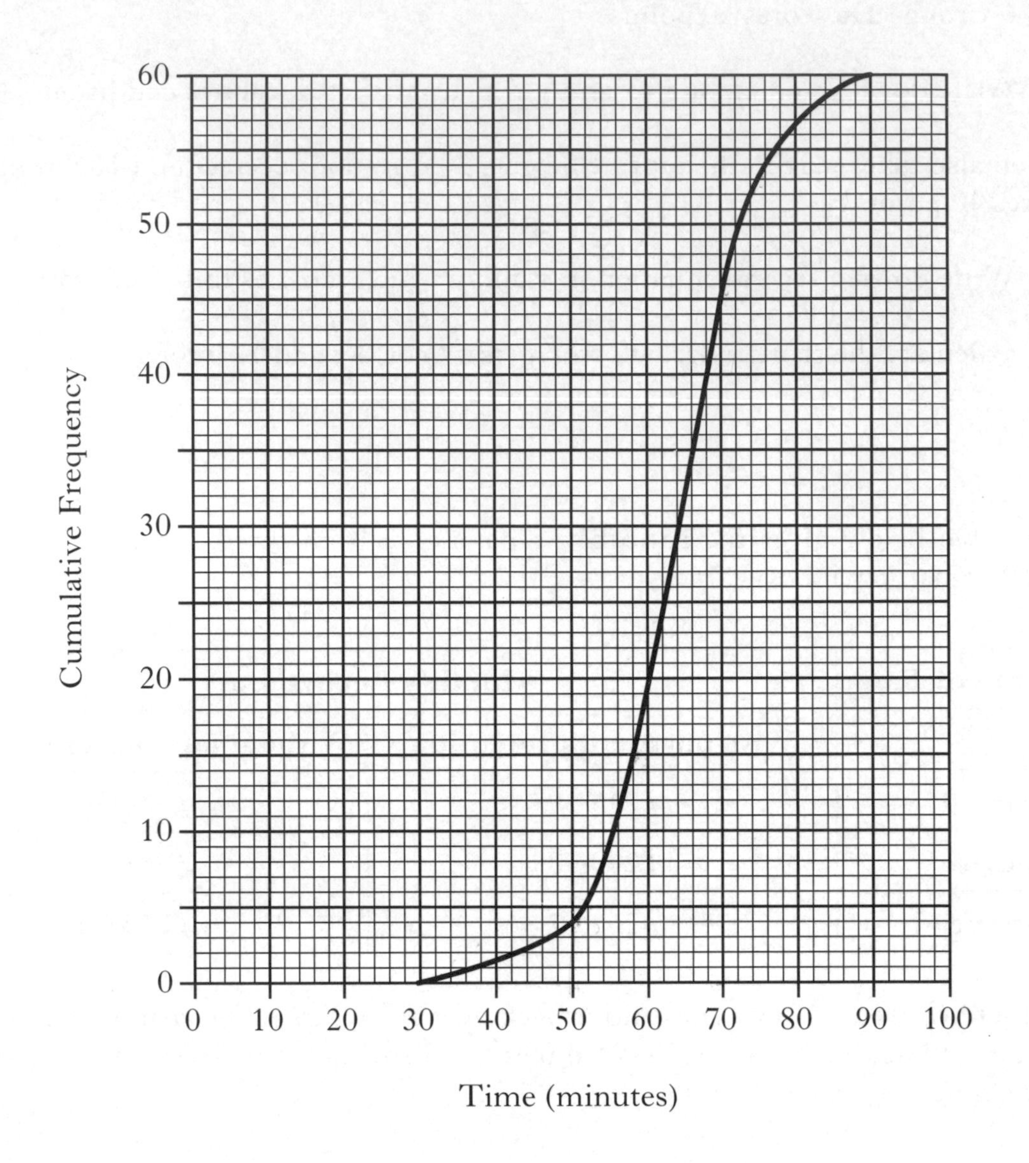

(*a*) How many runners completed the course in 50 minutes or less? **1**

(*b*) Calculate the semi-interquartile range for the data represented in the diagram. **3**

Marks

9. Jack works a basic week of 35 hours.

Any overtime is paid at time and a half.

One week he works for 39 hours and is paid £255·84.

How much is he paid for each hour of **overtime** that he works? **3**

[Turn over

Marks

10. Seamus has been offered jobs by both Paywell and Highpay. He constructs a spreadsheet to allow him to compare the salaries he has been offered. Part of the spreadsheet is shown below.

	A	B	C	D	E
1			**Paywell**		
2		**Basic salary**	**Bonus**	**Annual gross salary**	**Total earned to date**
3					
4	**Year 1**	£15,000	£1,250	£16,250	£16,250
5	**Year 2**	£15,600	£1,300	£16,900	£33,150
6	**Year 3**	£16,200	£1,350	£17,550	
7	**Year 4**	£16,800	£1,400		
8	**Year 5**	£17,400			
9					
10			**Highpay**		
11		**Basic salary**	**Bonus**	**Annual gross salary**	**Total earned to date**
12					
13	**Year 1**	£12,000	£1,200	£13,200	£13,200
14	**Year 2**	£14,000	£1,400	£15,400	£28,600
15	**Year 3**	£16,000	£1,600	£17,600	
16	**Year 4**	£18,000	£1,800		
17	**Year 5**	£20,000	£2,000		

Paywell offers an initial basic salary of £15 000, with a rise of £600 per annum and a bonus of one month's salary.

Highpay offers an initial basic salary of £12 000, with a rise of £2000 per annum and a bonus of 10% of his annual salary.

(*a*) Write down the **formula** to enter in cell C4 the bonus for Year 1. **1**

(*b*) Write down the **formula** to enter in cell E8 the total salary earned after 5 years with Paywell. **1**

(*c*) What will appear in cell E8? **2**

(*d*) Seamus intends to stay with the company for only 3 years.

Which company will allow him to earn more money in that time? **2**

Marks

11.

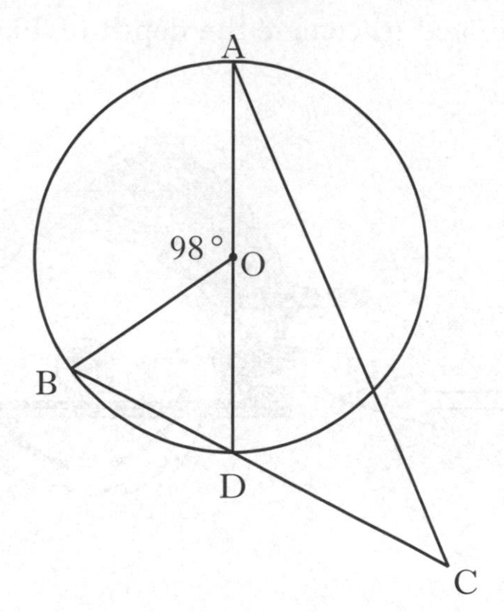

AD is a diameter of a circle, centre O.
B is a point on the circumference of the circle.
The chord BD is extended to a point C, outside the circle.
Angle BOA = 98 °.
DC = 9 centimetres. The radius of the circle is 7 centimetres.

Calculate the length of AC. **5**

[Turn over for Question 12 on *Page ten*

Marks

12. A circular saw can be adjusted to change the depth of blade that is exposed below the horizontal guide.

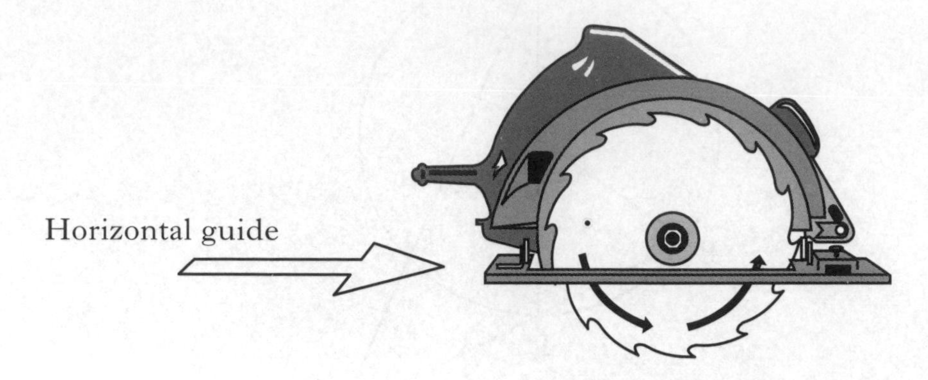

The circle, centre O, below represents the blade and the line AB represents part of the horizontal guide.

This blade has a radius of 110 millimetres.

If AB has length 140 millimetres, calculate the depth, d millimetres, of saw exposed.

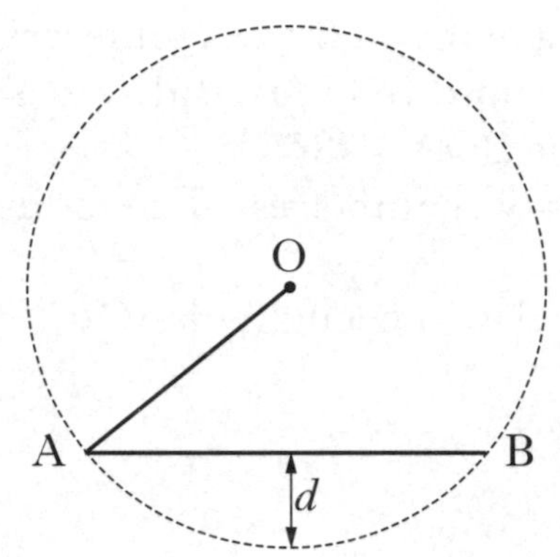

4

[*END OF QUESTION PAPER*]

INTERMEDIATE 2

2012

[BLANK PAGE]

X101/11/01

NATIONAL QUALIFICATIONS 2012

MONDAY, 21 MAY
9.00 AM – 9.45 AM

MATHEMATICS
INTERMEDIATE 2
Units 1, 2 and
Applications of Mathematics
Paper 1
(Non-calculator)

Read carefully

1 **You may NOT use a calculator.**

2 Full credit will be given only where the solution contains appropriate working.

3 Square-ruled paper is provided. If you make use of this, you should write your name on it clearly and put it inside your answer booklet.

FORMULAE LIST

Sine rule: $\frac{a}{\sin A} = \frac{b}{\sin B} = \frac{c}{\sin C}$

Cosine rule: $a^2 = b^2 + c^2 - 2bc \cos A$ or $\cos A = \frac{b^2 + c^2 - a^2}{2bc}$

Area of a triangle: $\text{Area} = \frac{1}{2}ab \sin C$

Volume of a sphere: $\text{Volume} = \frac{4}{3}\pi r^3$

Volume of a cone: $\text{Volume} = \frac{1}{3}\pi r^2 h$

Volume of a cylinder: $\text{Volume} = \pi r^2 h$

Standard deviation: $s = \sqrt{\frac{\sum (x - \bar{x})^2}{n-1}} = \sqrt{\frac{\sum x^2 - (\sum x)^2 / n}{n-1}}$, where n is the sample size.

ALL questions should be attempted. *Marks*

1. The National Debt of the United Kingdom was recently calculated as

£1 157 818 887 139.

Round this amount to four significant figures. 1

2. A teacher recorded the marks, out of ten, of a group of pupils for a spelling test.

Mark	Frequency
5	2
6	5
7	6
8	11
9	9
10	2

(*a*) Copy the frequency table and add a cumulative frequency column. 1

(*b*) For this data, find:

(i) the median; 1

(ii) the lower quartile; 1

(iii) the upper quartile. 1

(*c*) Draw a boxplot to illustrate this data. 2

[Turn over

Marks

3. The straight line with equation $4x + 3y = 36$ cuts the y-axis at A.

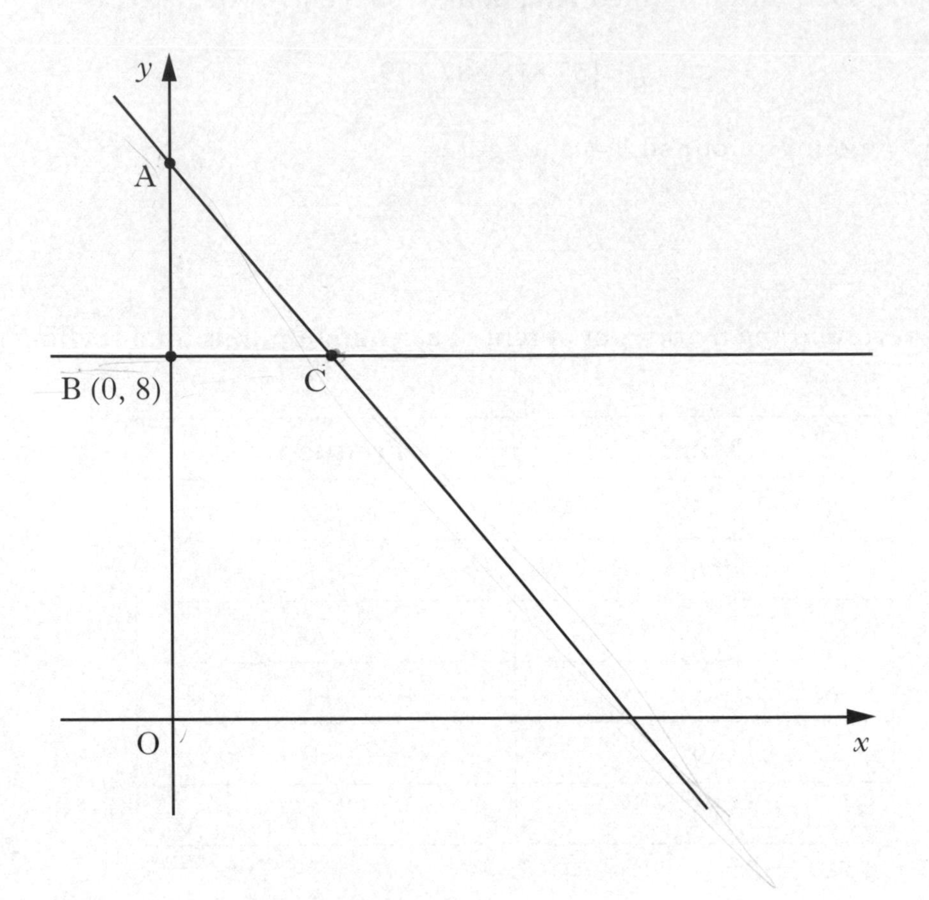

(*a*) Find the coordinates of A. **1**

This line meets the line through B (0, 8), parallel to the x-axis, at C as shown above.

(*b*) Find the coordinates of C. **2**

Marks

4.

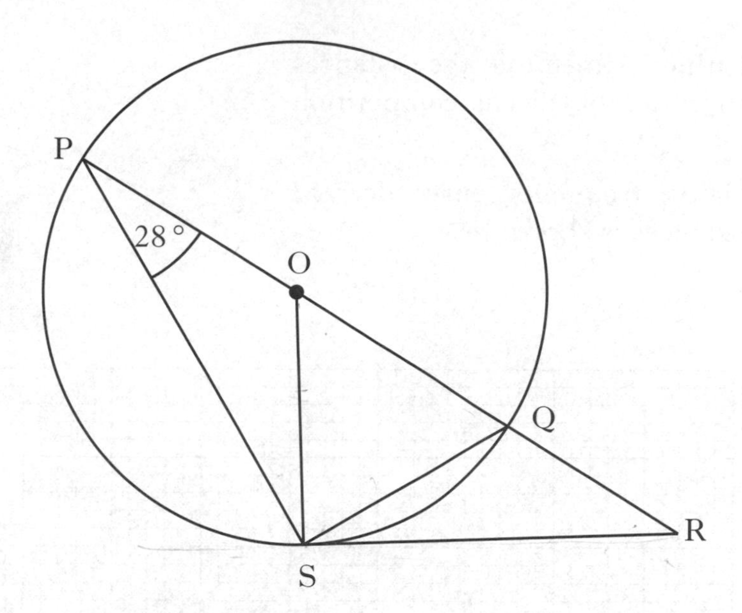

In the above diagram,

- O is the centre of the circle
- PQ is a diameter of the circle
- PQR is a straight line
- RS is a tangent to the circle at S
- angle OPS is 28°.

Calculate the size of angle QRS. **3**

5. One weekend, the attendances at five Premier League football matches were recorded.

8 900 12 700 59 200 10 300 9 700

The median attendance is 10 300.

(*a*) Calculate the mean attendance. **1**

(*b*) Which of the two "averages" – the mean or the median – is more representative of the data?

You must explain your answer. **1**

[Turn over

Marks

6. During an athletics meeting, the distances of 80 attempts in the discus competition are recorded.

The cumulative frequency curve derived from the distances is shown below.

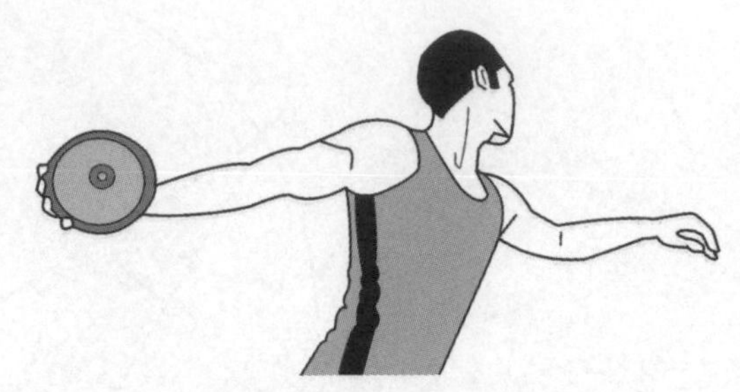

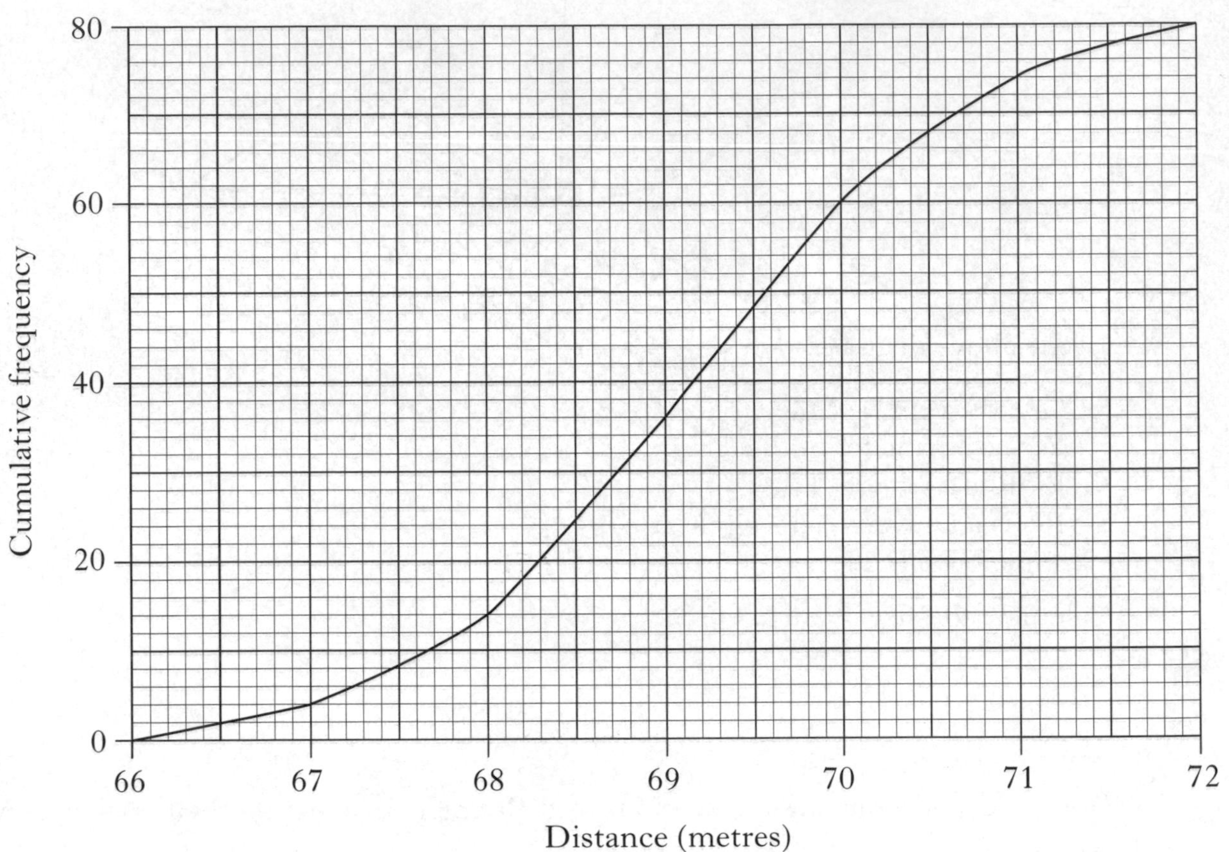

Use the curve to find the interquartile range of the distances. **3**

Marks

7.

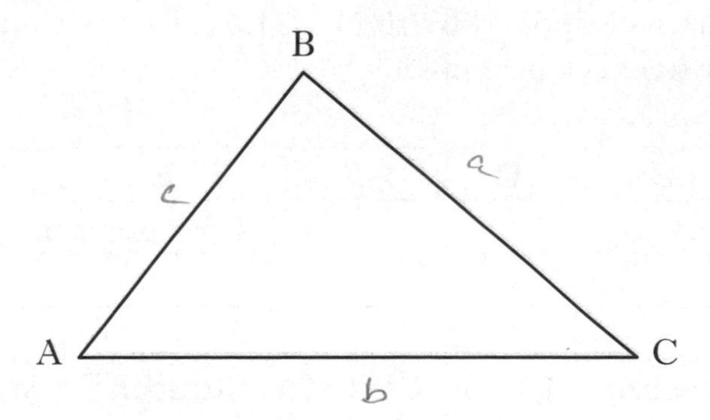

The area of triangle ABC is 20 square centimetres.

AC = 16 centimetres and $\sin C = \frac{1}{4}$.

Calculate the length of BC. **2**

8. (*a*) Factorise

$$a^2 + 2ab + b^2.$$ **1**

(*b*) Hence, or otherwise, find the value of

$$94^2 + 2 \times 94 \times 6 + 6^2.$$ **2**

[Turn over

Marks

9. Maureen has her electricity supplied by the Use Less Power Company. She has designed a spreadsheet to check her bills.

	A	B	C	D	E	F	G	H	I
1	**Use Less Power Company**					**Cost per unit = 16p**			
2									
3									
4		**Previous Reading**	**Present Reading**	**Units Used**	**Cost of Units**	**Standing Charge**	**Sub-total**	**VAT at 5%**	**Total cost**
5									
6	**Jan–Mar**	75 812	76 915	1103	£176·48	£14·99	£191·47	£9·57	£201·04
7	**Apr–Jun**	76 915	77 408	493	£78·88	£14·99	£93·87	£4·69	£98·56
8	**Jul–Sep**	77 408	77 632	224	£35·84	£14·99	£50·83		
9	**Oct–Dec**	77 632	78 519	887					

She receives a bill each quarter. Electricity costs 16p per unit and there is a standing charge of £14·99 per quarter.

(*a*) Write down the **formula** to enter in cell E8 the cost of the units for the period from July to September. **1**

(*b*) Write down the **formula** to enter in cell H8 the cost of the VAT at 5% for the period from July to September. **1**

(*c*) What value will appear in cell I8? **2**

Marks

10. A copy of Logan Pollock's payslip is shown below for one week in February.

Name L. Pollock	**Employee No.** 027	**Tax Code** 64L	**Week Ending** 14/02/2012
Basic Pay £296·00	**Overtime Pay** £55·50	**Bonus** —	**Gross Pay** £351·50
National Insurance £20·04	**Income Tax** £45·40	**Pension** £21·09	**Deductions** £86·53
			Net Pay £264·97

Logan worked 40 hours for his basic pay.

If overtime was paid at the rate of "time and a half", calculate how many hours of overtime he worked during that week. **3**

[*END OF QUESTION PAPER*]

[BLANK PAGE]

X101/11/02

NATIONAL QUALIFICATIONS 2012

MONDAY, 21 MAY
10.05 AM – 11.35 AM

MATHEMATICS
INTERMEDIATE 2
Units 1, 2 and
Applications of Mathematics
Paper 2

Read carefully

1 **Calculators may be used in this paper.**

2 Full credit will be given only where the solution contains appropriate working.

3 Square-ruled paper is provided. If you make use of this, you should write your name on it clearly and put it inside your answer booklet.

FORMULAE LIST

Sine rule: $\frac{a}{\sin A} = \frac{b}{\sin B} = \frac{c}{\sin C}$

Cosine rule: $a^2 = b^2 + c^2 - 2bc \cos A$ or $\cos A = \frac{b^2 + c^2 - a^2}{2bc}$

Area of a triangle: $\text{Area} = \frac{1}{2}ab \sin C$

Volume of a sphere: $\text{Volume} = \frac{4}{3}\pi r^3$

Volume of a cone: $\text{Volume} = \frac{1}{3}\pi r^2 h$

Volume of a cylinder: $\text{Volume} = \pi r^2 h$

Standard deviation: $s = \sqrt{\frac{\sum (x - \bar{x})^2}{n-1}} = \sqrt{\frac{\sum x^2 - (\sum x)^2 / n}{n-1}}$, where n is the sample size.

ALL questions should be attempted. *Marks*

1. The diagram below shows a circle, centre C.

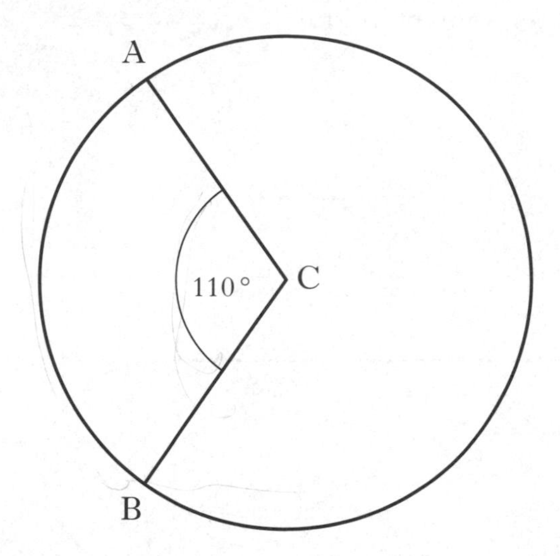

The circumference of the circle is 40·8 centimetres.

Calculate the length of the minor arc AB. **2**

2. Multiply out the brackets and collect like terms.

$(3x - 5)(x^2 + 2x - 6)$ **3**

[Turn over

Marks

3. A health food shop produces cod liver oil capsules for its customers.

Each capsule is in the shape of a cylinder with hemispherical ends as shown in the diagram below.

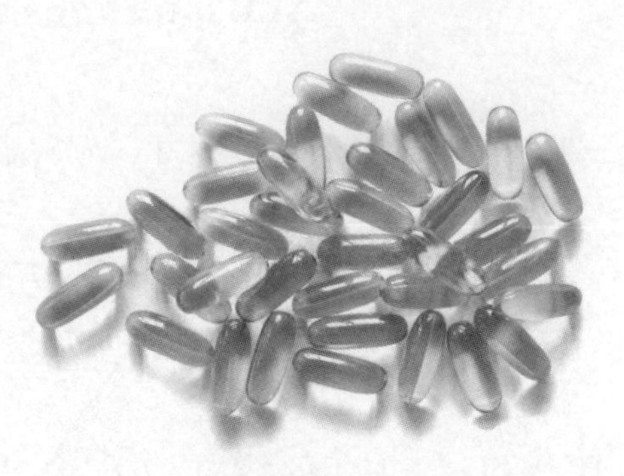

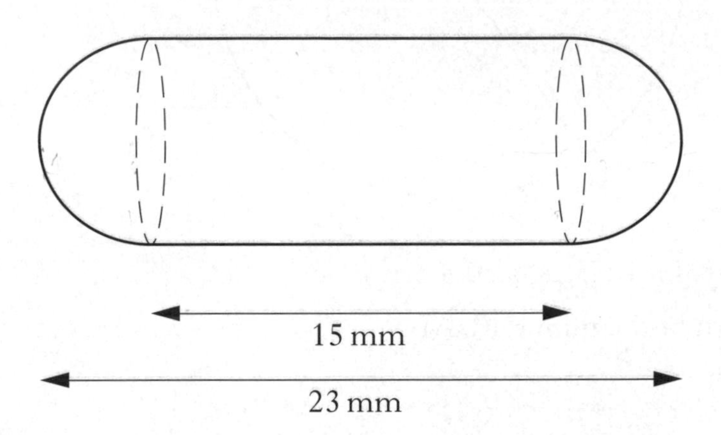

The total length of the capsule is 23 millimetres and the length of the cylinder is 15 millimetres.

Calculate the volume of one cod liver oil capsule. **4**

Marks

4. Stationery Systems offers a photocopying service to its customers. The flowchart below shows how charges are calculated for any number of copies.

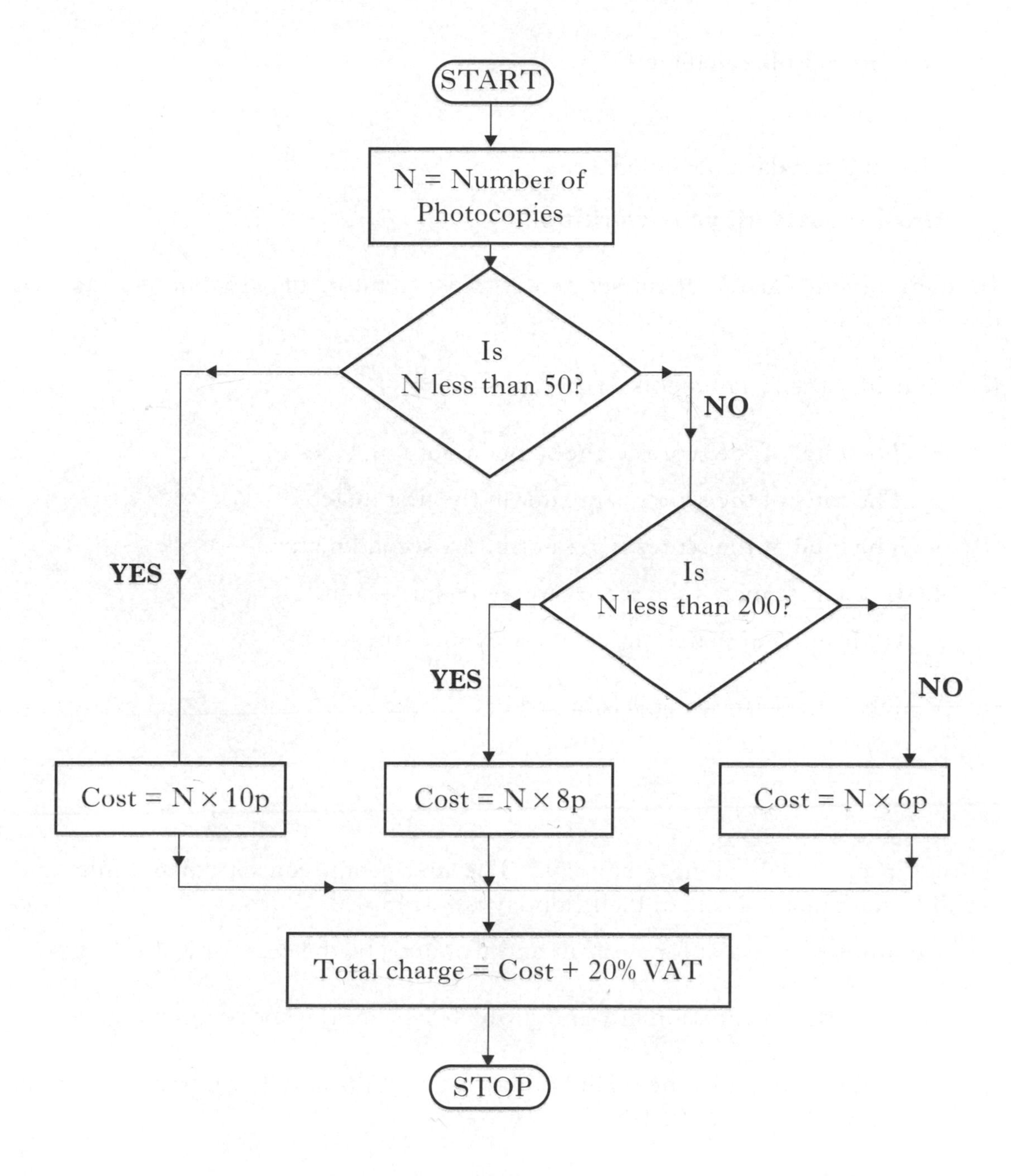

Use the flowchart to calculate the total charge for Kamran who makes 360 photocopies. **4**

[Turn over

Marks

5. A ten-pin bowling team recorded the following six scores in a match.

134 102 127 98 104 131

(*a*) For this sample calculate:

(i) the mean;

(ii) the standard deviation.

Show clearly all your working. 4

In their second match their six scores have a mean of 116 and a standard deviation of 12·2.

(*b*) Consider the 5 statements written below.

1 The total of the scores is the same in both matches.

2 The total of the scores is greater in the first match.

3 The total of the scores is greater in the second match.

4 In the first match the scores are more spread out.

5 In the second match the scores are more spread out.

Which of these statements is/are true? 2

6. Three groups are booking a holiday. The first group consists of 6 adults and 2 children. The total cost of their holiday is £3148.

Let x pounds be the cost for an adult and y pounds be the cost for a child.

(*a*) Write down an equation in x and y which satisfies the above information. 1

The second group books the same holiday for 5 adults and 3 children. The total cost of their holiday is £3022.

(*b*) Write down a second equation in x and y which satisfies this information. 1

(*c*) The third group books the same holiday for 2 adults and 4 children. The travel agent calculates that the total cost is £2056.

Has this group been overcharged?

Justify your answer. 4

Marks

7. A network diagram is shown below.

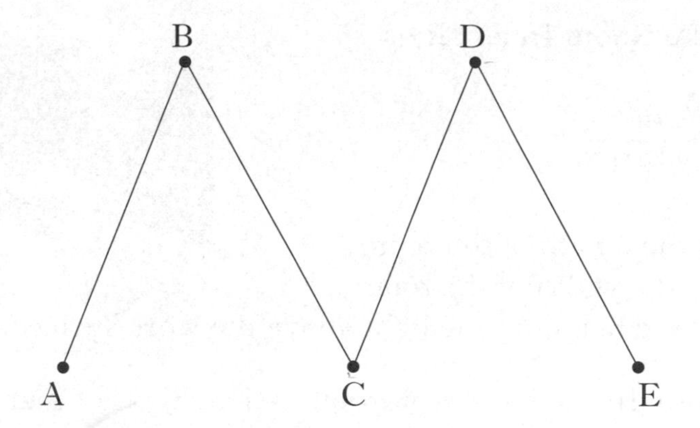

Copy the diagram and add one arc so that all the nodes are even. **1**

8. The Bank of Salamander offers loans to its customers.

The table shown below can be used to calculate loan repayments.

		60 months	**48 months**	**24 months**
		Monthly repayment (£)	Monthly repayment (£)	Monthly repayment (£)
With payment protection	£20 000	467·85	555·43	998·23
	£15 000	351·89	417·57	749·67
	£7500	177·94	210·79	376·84
Without payment protection	£20 000	388·65	471·72	888·47
	£15 000	292·49	354·79	667·35
	£7500	148·29	179·40	335·68

Amy requires to borrow £15 000 to buy a car.

How much will the loan cost her if she repays it over 24 months, **without payment protection**? **3**

[Turn over

Marks

9. The Room Index is used to calculate the amount of light needed in a workroom.

The formula for the Room Index, R, is

$$R = \frac{LW}{H(L+W)}$$

where L metres is the length of the room,
W metres is the width of the room
and H metres is the height of the light above the work surface.

Calculate the Room Index for a workroom 4·4 metres long and 3·2 metres wide with the light 1·4 metres above the work surface. **3**

10. A tanker delivers oil to garages.

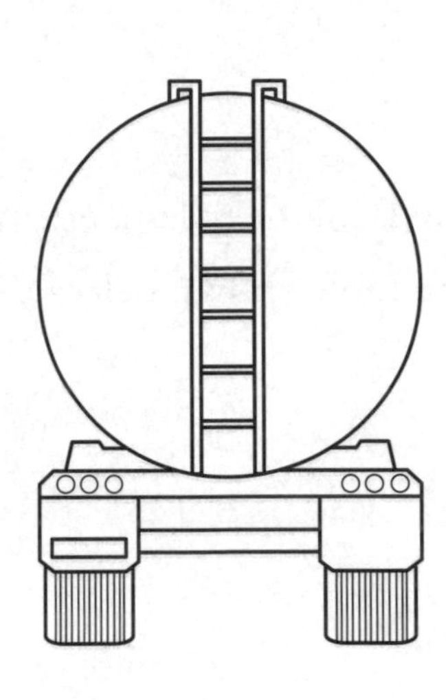

The tank has a circular cross-section as shown in the diagram below.

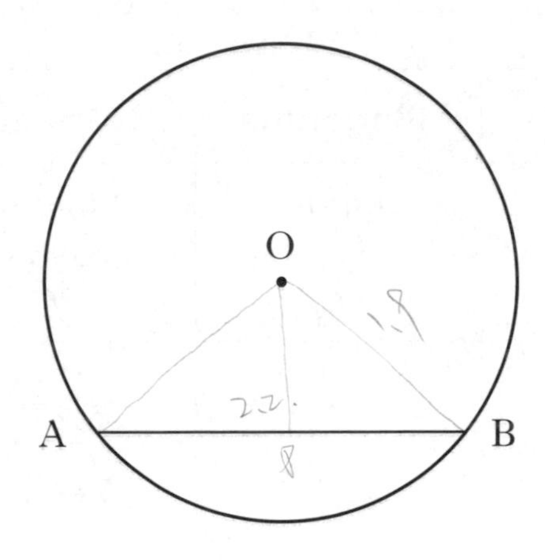

Depth of oil

The radius of the circle, centre O, is 1·9 metres.

The width of the surface of the oil, represented by AB in the diagram, is 2·2 metres.

Calculate the depth of the oil in the tanker. **4**

Marks

11. A dental practice keeps a record of the number of patients visiting the surgery over a period of time.

The information is shown below.

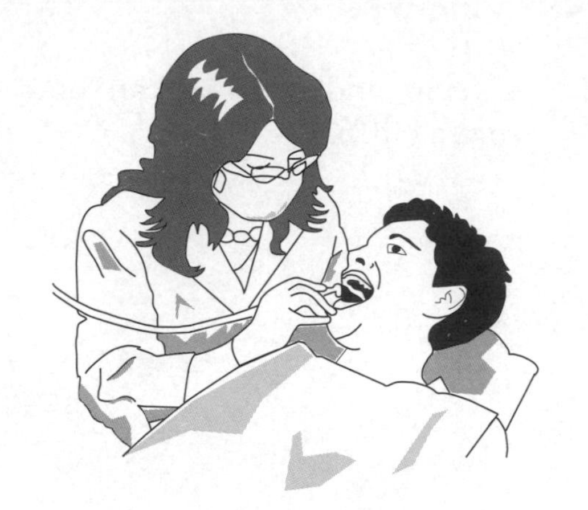

Number of patients	*Number of days*
6 – 10	4
11 – 15	8
16 – 20	10
21 – 25	18
26 – 30	7
31 – 35	3

Taking the number of patients to be at the mid-point of each interval, calculate the mean number of patients visiting the surgery per day. **5**

[Turn over

Marks

12. A yacht and a canoe can be seen from a clifftop.

In the diagram below, Y and C represent the positions of the yacht and the canoe.

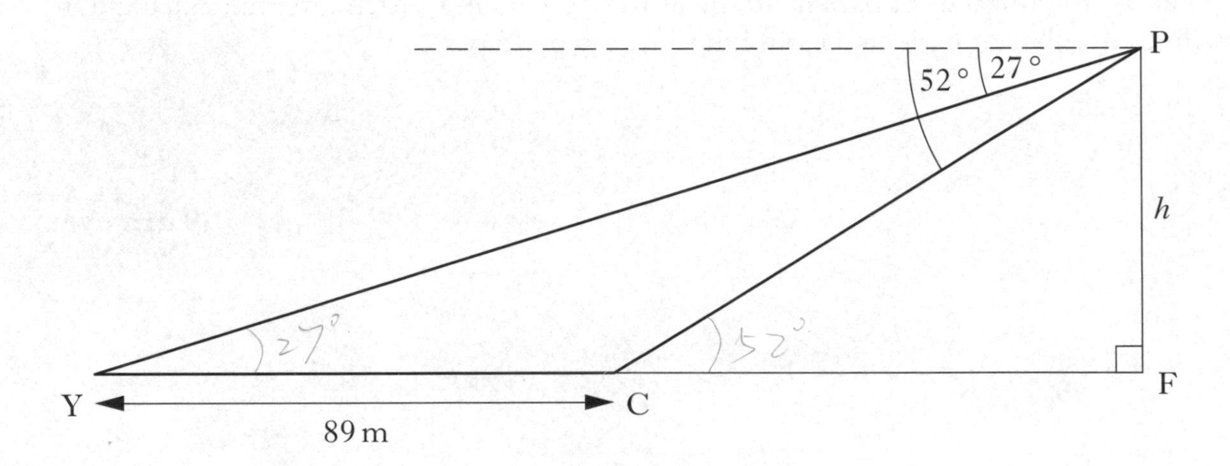

From a point P on the clifftop:

- the angle of depression of the yacht is 27°;
- the angle of depression of the canoe is 52°.

The distance between the yacht and the canoe is 89 metres.

Calculate the height, h, metres, of the cliff. **5**

Marks

13. Due to the threat of global warming, scientists recommended in 2010 that the emissions of greenhouse gases should be reduced by 50% by the year 2050.

The government decided to reduce the emissions of greenhouse gases by 15% **every ten years**, starting in the year 2010.

Will the scientists' recommendations have been achieved by 2050?

You must give a reason for your answer. **4**

[*END OF QUESTION PAPER*]

[BLANK PAGE]

INTERMEDIATE 2 | ANSWER SECTION

BrightRED ANSWER SECTION FOR

SQA INTERMEDIATE 2 MATHEMATICS: UNITS 1, 2 and APPLICATIONS 2008–2012

MATHEMATICS INTERMEDIATE 2 UNITS 1, 2 AND APPLICATIONS PAPER 1 2008 (NON-CALCULATOR)

1. gradient is 4

2. $3x^2 - 5x - 10$

3. (*a*) 12th

(*b*) $\frac{5}{20}$ or equivalent

4. (*a*) $(x + y)(x - y)$

(*b*) 86

5. (*a*)

Number of books	*Frequency*	*Cumulative Frequency*
0	1	1
1	2	3
2	3	6
3	5	11
4	5	16
5	6	22
6	2	24
7	1	25

(*b*) $Q_2 = 4$, $Q_1 = 2{\cdot}5$, $Q_3 = 5$

(*c*) 1·25

(*d*) number of textbooks more spread out for girls

6. 40 sq cm

7. 19°

8. T, U

9. £3600

10. (*a*)

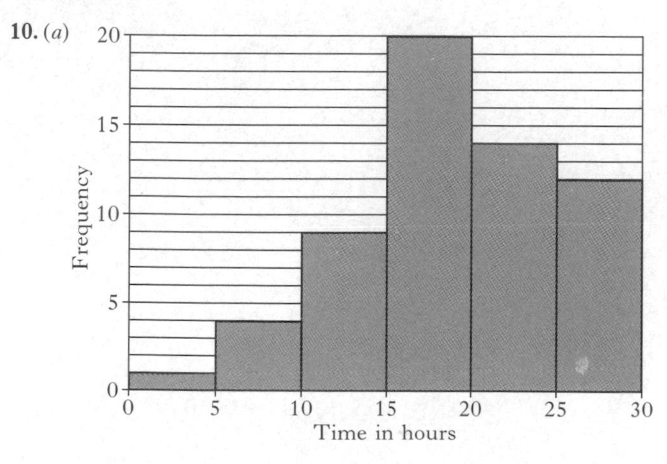

(*b*) 18 hours

(*c*) (On average) students spend mcre time watching television than studying

11. 45

MATHEMATICS INTERMEDIATE 2 UNITS 1, 2 AND APPLICATIONS PAPER 2 2008

1. £9625·93

2. (*a*) 58 600 cubic cm

(*b*) 29·9 cm

3. (*a*) 14·8

(*b*) The physics marks were more consistent than the maths marks (since $6{\cdot}8 < 14{\cdot}8$)

(*c*) $y = \frac{1}{2}x + 20$

(*d*) 58%

4. (*a*) $280x + 70y = 5250$

(*b*) $210x + 40y = 3800$

(*c*) Calls cost 16 pence per minute, texts cost 11 pence each

5. Angle EDF = 111·8°

6. (*a*) A = £75·00
B = £1·20
C = £155·07

(*b*) £5

7. (*a*) = C22*1·004

(*b*) = B23 + 250

(*c*) £13 557·59

8. £ 1976·40

9. 16 cm

10. £5·20

MATHEMATICS INTERMEDIATE 2 UNITS 1, 2 AND APPLICATIONS PAPER 1 2009 (NON-CALCULATOR)

1. (*a*)

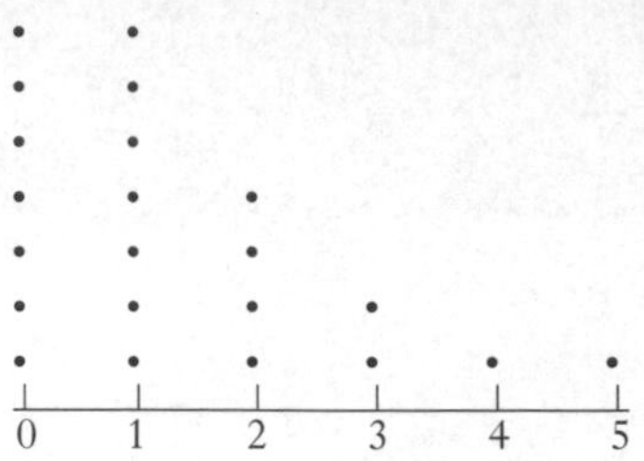

0 1 2 3 4 5

(*b*) A

2. $y = 3x - 1$

3. $(x - 8)(x + 3)$

4. $2x^3 + 7x^2 - 16x - 5$

5. (*a*) (i) 58·5 (ii) 11

(*b*) In December, the marks (on average) are better and less spread out

6. Any value for a such that $270 < a < 360$.

7. −1

8.

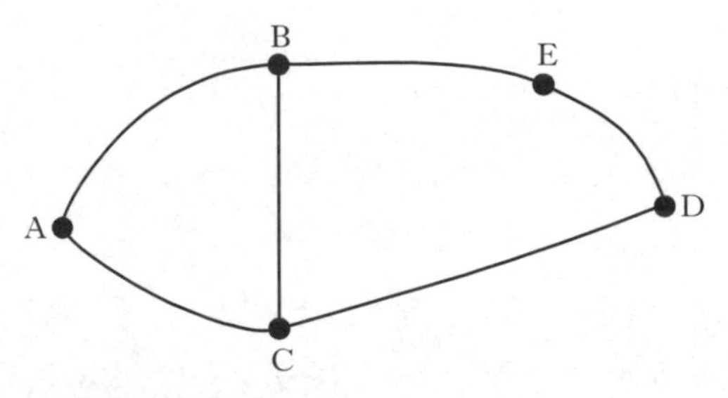

9. (*a*) =SUM(E6:E9)

(*b*) =D6*(1+B2/100)

10. (*a*) S = 700cm^2

(*b*) a = 9cm

MATHEMATICS INTERMEDIATE 2 UNITS 1, 2 AND APPLICATIONS PAPER 2 2009

1. There were 3 sales fewer in 2008 or There were fewer sales in 2008 because $2997 < 3000$

2. (*a*) 172 cm
(*b*) 4·8 cm

3. 882 000 mm^3

4. (*a*) $14x + 60y = 344{\cdot}30$
(*b*) $21x + 40y = 368{\cdot}95$
(*c*) A car costs £11·95 and a passenger £2·95

5. 313 square inches

6. 68·6°

7. £4851·36

8. £100

9. £642·50

10. 8·6 metres

11. 3·14 metres

12. 105 pence (or £1·05)

MATHEMATICS INTERMEDIATE 2 UNITS 1, 2 AND APPLICATIONS PAPER 1 2010 (NON-CALCULATOR)

1. $y = -\frac{4}{3}x + 8$

2. (*a*)

Shoe size	frequency	cumulative frequency
5	3	3
6	4	7
7	7	14
8	3	17
9	2	19
10	0	19
11	1	20

(*b*) (i) 7 (ii) 6 (iii) 8

(*c*)

5 6 7 8 9 10 11

3. 113·04 cubic centimetres

4. (*a*) $(x + 3)(x - 2)$

(*b*) $3x^3 + 17x^2 + 7x - 2$

5. Yes, with valid explanation e.g. Yes, because the route has only 2 odd vertices – museum and church.

6. 8 centimetres

7. (*a*) $a = 144$

(*b*) $n = 9$

8. £1952

MATHEMATICS INTERMEDIATE 2 UNITS 1, 2 AND APPLICATIONS PAPER 2 2010

1. £155 000

2. 150°, 200°, 10°

3. £11

4. (*a*) (i) 7 (ii) 3·958

(*b*) The team scores more points under the new coach. The team is more consistent.

5. $x = 7, y = -2$

6. No, because it will take 23 minutes to tidy.

7. (*a*) = C6 − B6

(*b*) 12

8. £58 (± 0·30)

9. 1342·35 square centimetres

10. Proof

$(x + 7)(x + 3)$

evidence of four correct terms

$x^2 + 7x + 3x + 21$ leading to

$x^2 + 10x + 21$

11. 25·3 centimetres

12. 126·5 metres

13. 3·45 metres

14. 3 hours

15. (*a*) cumulative frequency curve

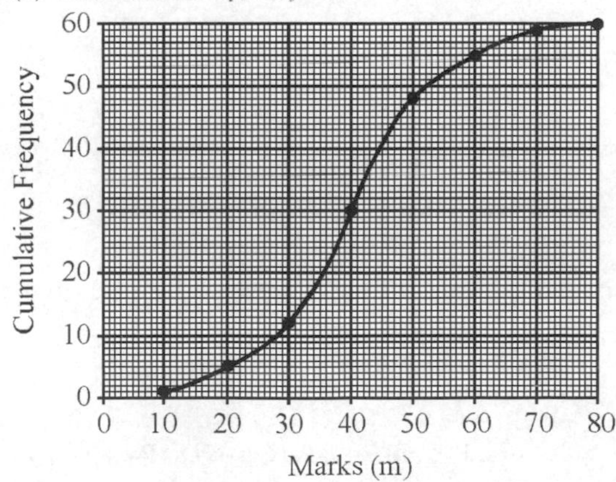

(*b*) (i) 32 (ii) 48

(*c*) 8

MATHEMATICS INTERMEDIATE 2 UNITS 1, 2 AND APPLICATIONS PAPER 1 2011 (NON-CALCULATOR)

1. (*a*) (i) $Q_2 = 6{\cdot}5$
(ii) $Q_1 = 5$
(iii) $Q_3 = 9$

(*b*)

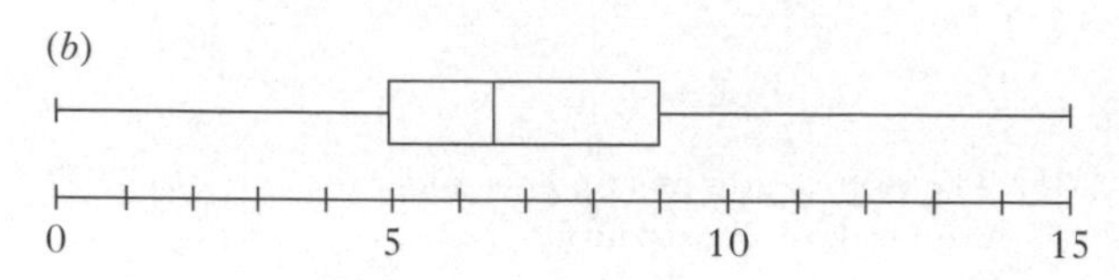

(*c*) The trains are not as late as the buses **or** the trains are more reliable

2. $(x - 7)(x + 3)$

3. $6x^2 - 12x - 14$

4. 138°

5. 25 metres

6. £235·13 **or** £235·12

7. To prove $\cos B = \frac{5}{9}$

$$\cos B = \frac{a^2 + c^2 - b^2}{2\,a\,c} \text{ (using cosine rule)}$$
$$= \frac{6^2 + 3^2 - 5^2}{2 \times 6 \times 3}$$
$$= \frac{36 + 9 - 25}{36}$$
$$= \frac{20}{36}$$
$$= \frac{5}{9}$$

8.

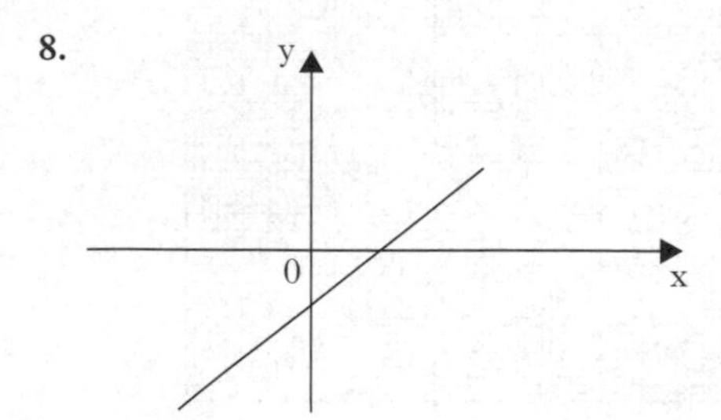

9. (*a*) P-G-A-N-E, P-G-E-A-N, P-G-E-N-A, P-G-N-A-E, P-G-N-E-A

A tree diagram showing the following routes: P-G-A-N-E, P-G-E-AN, P-G-E-N-A, P-G-N-A-E, P-G-N-E-A

(*b*) Shortest distance, finishing at Newcastle, is 431 miles

10. $\frac{4}{5}$

MATHEMATICS INTERMEDIATE 2 UNITS 1, 2 AND APPLICATIONS PAPER 2 2011

1. −9/10

2. £147 900

3. (*a*) 106 cubic metres

(*b*) 17·4 metres

4. 25·1 square metres

5. (*a*) (i) $\bar{x} = 41$
(ii) $s = 2{\cdot}1$

(*b*) Yes, with reasons covering both conditions

6. (*a*) $24x + 6y = 60$

(*b*) $20x + 10y = 40$

(*c*) 25 points

7. Finesave without payment protection

8. (*a*) 4 runners
(*b*) 6

9. £9·36

10. (*a*) = B4/12

(*b*) = SUM(D4:D8)

(*c*) £87 750

(*d*) £50 700, £46 200, so Paywell pays more

11. 21 centimetres

12. 25·1 millimetres

MATHEMATICS INTERMEDIATE 2 UNITS 1, 2 AND APPLICATIONS PAPER 1 2012 (NON-CALCULATOR)

1. £1 158 000 000 000

2. (*a*)

mark	frequency	cumulative frequency
5	2	2
6	5	7
7	6	13
8	11	24
9	9	33
10	2	35

(*b*) (i) 8
(ii) 7
(iii) 9

(*c*)

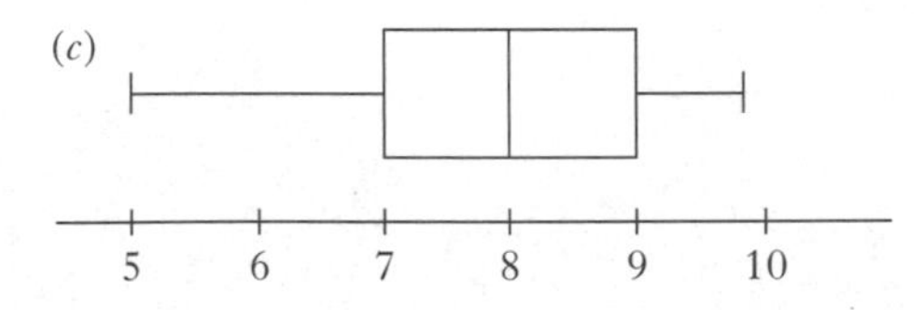

3. (*a*) A(0, 12)

(*b*) C(3, 8)

4. 34°

5. (*a*) 20 160

(*b*) The median, with reason. The reason must refer to the fact that the mean is affected by one very high attendance or that the median is closer to the majority of the attendances.

6. 1·7

7. 10 centimetres

8. (*a*) $(a + b)^2$

(*b*) 10 000

9. (*a*) =D8*0·16

(*b*) =G8*0·05

(*c*) £53·37

10. 5 hours

MATHEMATICS INTERMEDIATE 2 UNITS 1, 2 AND APPLICATIONS PAPER 2 2012

1. 12.5 centimetres

2. $3x^3 + x^2 - 28x + 30$

3. 1022 mm^3

4. £25·92

5. (*a*) (i) 116
(ii) 16·33

(*b*) 1 and 4 (The total score is the same in both matches and in the first match the scores are more spread out.)

6. (*a*) $6x + 2y = 3148$

(*b*) $5x + 3y = 3022$

(*c*) Yes. The group has been overcharged by £10.

7.

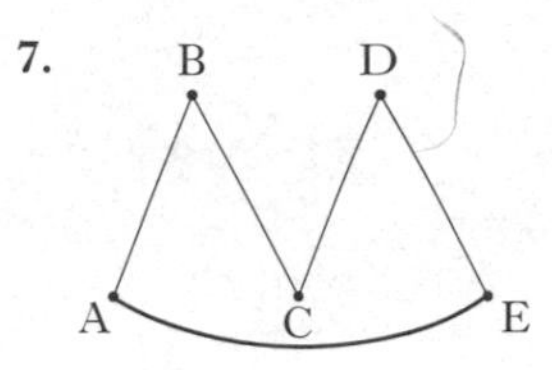

8. £1016·40

9. 1·32

10. 0·4

11. 20·5

12. 75·3 metres

13. No, 0·522 > 0·5

Published by Bright Red Publishing Ltd, 6 Stafford Street, Edinburgh, EH3 7AU
Tel: 0131 220 5804, Fax: 0131 220 6710, enquiries: sales@brightredpublishing.co.uk,
www.brightredpublishing.co.uk

Official SQA answers to 978-1-84948-276-9
2008-2012